SIDE by SIDE TV

VIDEO WORKBOOK 1A

Steven J. Molinsky
Bill Bliss

Contributing Authors
Elizabeth Handley
Judy Boyle

D1292156

 PRENTICE HALL REGENTS, Englewood Cliffs, New Jersey 07632

VIDEO WORKBOOK 1A

Publisher: Tina Carver
Editorial Production/Design Manager: Dominick Mosco
Manufacturing Buyer: Ray Keating
Project Manager: Harriet Dishman
Design and Composition: PC&F, Inc.
Video Stills: Elizabeth Gallagher

Illustrations: Richard E. Hill

Clock times in this workbook enable users to easily find the corresponding video material. Page numbers that appear on the videos refer to pages in the *Side by Side* textbooks, which can also be used as part of this video-based language learning program.

© 1995 by Prentice Hall Regents
Prentice-Hall, Inc.
A Simon & Schuster Company
Englewood Cliffs, New Jersey 07632

In association with

Video Publishing Group, Inc.

Printed in the United States of America

10 9 8 7 6 5 4 3 2 1

ISBN 0-13-815127-X

Prentice-Hall International (UK) Limited, *London*
Prentice-Hall of Australia Pty. Limited, *Sydney*
Prentice-Hall Canada Inc., *Toronto*
Prentice-Hall Hispanoamericana, S.A., *Mexico*
Prentice-Hall of India Private Limited, *New Delhi*
Prentice-Hall of Japan, Inc., *Tokyo*
Simon and Schuster Asia Pte. Ltd., *Singapore*
Editora Prentice-Hall do Brasil Ltda., *Rio de Janeiro*

CONTENTS

• • • • • PREFACE • • • • •

Side by Side TV combines education and entertainment through a variety of comedy sketches, on-location interviews, rap numbers, and music videos—all designed to help learners of English build their language skills. This innovative program may be used in conjunction with the *Side by Side* textbook series, or on its own as an exciting "stand-alone" video-based English course.

Each level of *Side by Side TV* is divided into 13 segments. Within each segment there are short scenes that highlight important structures, functions, and vocabulary items.

The *Side by Side TV Video Workbooks* are designed to be used with the videos as a learning companion. Corresponding to each level of the video, they offer an outstanding variety of exercises and activities to promote language development. The *Video Workbooks* can be used in class or for self-study at home.

FEATURES OF THE VIDEO WORKBOOKS

- *Segment-Opening Pages* indicate the language focus of the segment, the scenes of the segment, and key vocabulary featured in that segment.

- *Exercises and Activities* are intended to help learners interact with the scenes in the video. Certain exercises and activities require use of the video, while others do not. Those that require use of the video are indicated with a small videocassette symbol. For each scene, the clock-time on the video is indicated on the workbook page.

  We encourage viewers to develop their own strategies for working with the exercises. Some viewers may wish to watch a scene one or more times before doing an exercise. Others may wish to complete an exercise while watching a scene. And others may wish to do an exercise before watching, as a way of predicting what might happen in the scene.

- *Scripts* are provided at the end of each workbook segment. Viewers can read along as they watch, read before they watch to preview a scene, or read later for review and practice.

- A *Summary Page* highlights the grammar structures and functions featured in that segment.

The mission of *Side by Side TV* is to provide learners of English with an exciting, motivating, and enjoyable language learning experience through television. We hope that *Side by Side TV* and the accompanying *Video Workbooks* offer viewers a language learning experience that is dynamic, interactive, . . . and fun!

Steven J. Molinsky
Bill Bliss

SEGMENT 1

- **Personal Information**
- **To Be: Introduction**

"Start a conversation with some basic information

. . . talking Side by Side."

PROGRAM LISTINGS

SBS-TV Backstage Bulletin Board

TO: Production Crew
Sets and props for this segment:

Classroom
chalkboard
desks

Hospital
counter
telephone
stretcher

TO: Cast Members
Key words in this segment:

name
address
phone number/
  telephone number
zip code
Social Security
  number

WHAT'S YOUR NAME?

SOUND CHECK

| address | from | name | phone number |

1 What's your ____name____? My ____name____ is Maria.

2 What's your _____? My _____ is 235 Main Street.

3 What's your _____? My _____ is 741-8906.

4 Where are you _____? I'm _____ Mexico City.

WHOSE LINE?

1 "I'm from Mexico City." Teacher (Maria)

2 "What's your address?" Teacher Maria

3 "My name is Maria." Teacher Maria

4 "My phone number is 741-8906." Teacher Maria

5 "What's your name?" Teacher Maria

CLOSE-UP

You're on Side by Side TV! Answer these questions.

1 What's your name? ....

2 What's your address? ....

3 What's your phone number? ....

4 Where are you from? ....

SOUND CHECK

Michael Pearson

Asako Fujiyama

Ron Walters

1 My address is ____.

   (a.) 9 Park Street

   b. 19 Park Street

2 My address is ____.

   a. 558 School Street

   b. 568 School Street

3 My address is ____.

   a. 739 River Street

   b. 749 River Street

4 My phone number is ____.

   a. 539-4271

   b. 839-4270

5 My phone number is ____.

   a. 285-6842

   b. 295-6843

6 My phone number is ____.

   a. 263-2739

   b. 263-2829

7 I'm from ____.

   a. Toledo

   b. Toronto

8 I'm from ____.

   a. Tokyo

   b. Kyoto

9 I'm from ____.

   a. Chicago

   b. Chattanooga

INTERVIEW

Interview two friends. Ask questions and write the answers in your reporter's notebook.

Name: ....

Address: ....

Phone Number: ....

From: ....

Name: ....

Address: ....

Phone Number: ....

From: ....

ON CAMERA

You're the nurse! Fill in the information about Mr. Chen.

Emergency Room
Personal Information Form

Name: ___William Chen_____

Address: _____River Street_____

_____Brooklyn_____New York_____
City   State   Zip Code

Telephone: _____

Social Security Number: _____

Now you're the patient! Fill in information about yourself.

Emergency Room
Personal Information Form

Name: _____

Address: _____

_____
City   State   Zip Code

Telephone: _____

Social Security Number: _____

EDITING MIX-UP

The video editor made a mistake! Put the nurse's lines in the correct order.

____ Is that in Brooklyn?

____ Mr. Chen, what's your zip code?

____ And what's your Social Security number?

_1_ Your name, please?

____ And your telephone number?

____ And what's your address, Mr. Chen?

Now write Mr. Chen's answers.

1. William Chen

2. _____

3. _____

4. _____

5. _____

6. _____

SOUND CHECK

| address | name | number | Social Security |
| telephone | Yes | zip code | |

1. Your ____name____, please?   William Chen.

2. And what's your _____, Mr. Chen?   694 River Street.

3. 694 River Street. Is that in Brooklyn?   _____.

4. Brooklyn . . . New York. And your _____ number?

   Mr. Chen. What's your telephone _____?   469-7750.

5. 469-7750. And what's your _____ number?   044-35-9862.

6. 044-35-9862. Tell me, Mr. Chen, what's your _____? . . .

   Mr. Chen? Mr. Chen? Hmm!

SCRAMBLED SOUND TRACK

The sound track is all mixed up. Put the words in the correct order.

1 name | your | What's | ?   *What's your name?*

2 ? | What's | number | your | telephone   _____

3 from | Where | ? | you | are   _____

4 address | My | is | Street | Main | 842 | .   _____

5 Security | number | 105-36-8954 | Social | is | My | .   _____

WHAT'S MY LINE?

1 What's you / **your** name?

2 What's / Where your address?

3 My / I'm from Toronto.

4 Where is / are you from?

5 And you / your telephone number?

6 What / What's your Social Security number?

ON CAMERA

You're on Side by Side TV! Interview these people.

1 _____ *What's your name?* _____

My name is Eduardo.

2 _____

19 Grand Avenue.

3 _____

235-9489.

4 _____

I'm from Seoul.

6

●●●●● **SEGMENT 1**

:09 WHAT'S YOUR NAME?

TEACHER:	What's your name?
MARIA:	My name is Maria.
TEACHER:	What's your address?
MARIA:	My address is 235 Main Street.
TEACHER:	What's your phone number?
MARIA:	My phone number is 741-8906.
TEACHER:	Where are you from?
MARIA:	I'm from Mexico City.

:41 SBS-TV ON LOCATION

INTERVIEWER:	What's your name?
MICHAEL:	My name is Michael Pearson.
ASAKO:	My name is Asako Fujiyama.
RON:	My name is Ron Walters. What's YOUR name?
INTERVIEWER:	What's your address?
MICHAEL:	My address is 9 Park Street.
ASAKO:	My address is 568 School Street.
RON:	My address is 749 River Street. What's YOUR address?
INTERVIEWER:	What's your phone number?
MICHAEL:	My phone number is 539-4271.
ASAKO:	My phone number is 295-6843.

RON:	My phone number is 263-2829. What's YOUR phone number?
INTERVIEWER:	Where are you from?
MICHAEL:	I'm from Toronto.
ASAKO:	I'm from Tokyo.
RON:	I'm from Chicago. Where are YOU from?

1:36 YOUR NAME, PLEASE?

NURSE:	Your name, please?
MR. CHEN:	William Chen.
NURSE:	William Chen. And what's your address, Mr. Chen?
MR. CHEN:	694 River Street.
NURSE:	694 River Street. Is that in Brooklyn?
MR. CHEN:	Yes.
NURSE:	Brooklyn . . . New York. And your telephone number?

(Mr. Chen groans. He doesn't hear the question.)

	Mr. Chen. What's your telephone number?
MR. CHEN:	469-7750. Oh!
NURSE:	469-7750. And what's your Social Security number?
MR. CHEN:	044-35-9862.

(As the nurse writes down his Social Security number, two attendants put Mr. Chen on a stretcher and take him away.)

NURSE:	044-35-9862. Tell me, Mr. Chen, what's your zip code? Mr. Chen? Mr. Chen? Hmm!

GRAMMAR

To Be

am	I'm from Mexico City. (I am)
is	What's your name? (What is)
	My name is Maria.
are	Where are you from?

FUNCTIONS

Asking for and Reporting Information

What's your name?
Your name, please?
   My name is *Maria.*

What's your address?
   My address is *235 Main Street.*

What's your phone number?
And your telephone number?
   My phone number is *741-8906.*

Where are you from?
   I'm from *Mexico City.*

What's your Social Security number?
   My Social Security number is *044-33-9862.*

What's your zip code?

Tell me, _____?

SEGMENT 2

- Rooms in the Home
- To Be: Subject Pronouns

"In the kitchen and the dining room, the attic, yard, and living room . . . we're Side by Side."

PROGRAM LISTINGS

SBS-TV Backstage Bulletin Board

TO: Production Crew
Sets and props for this segment:

Kitchen
table
chairs
newspaper
electric mixer
telephone

Yard
chairs
newspaper

Living Room
lamp
sofa
television

Office
desk
photograph
telephone

Bedroom
bed

TO: Cast Members
Key words in this segment:

basement
bathroom
bedroom
dining room

garage
kitchen
living room
yard

3:38 WHERE ARE YOU?

SETS AND SCENERY

Help the production crew with the sets for this segment.

| dining room | living room | kitchen | bedroom | yard | basement |

1 _____kitchen_____   2 _____   3 _____

4 _____   5 _____   6 _____

SOUND CHECK

1
a. I'm in the kitchen.
b. I'm in the living room.

2
a. We're in the dining room.
b. We're in the living room.

3
a. They're in the yard.
b. They're in the garage.

4:02 WHERE'S BOB?

SOUND CHECK

1
a. He's in the bedroom.
b. He's in the living room.

2
a. She's in the bedroom.
b. She's in the bathroom.

3
a. It's in the yard.
b. It's in the garage.

What's your favorite room at home?

1.   a. the dining room
      (b.) the living room

2.   a. the yard
      b. the dining room

3.   a. the kitchen
      b. the bedroom

4.   a. my bedroom
      b. my bathroom

5.   a. the bedroom
      b. the basement

CLOSE-UP

What's YOUR favorite room?

....

INTERVIEW

Interview five friends. Ask, "What's your favorite room?" Write your answers in your reporter's notebook.

Friend's Name	Favorite Room
1	
2	
3	
4	
5	

WHOSE LINE?

1	"Hi, Billy. This is Mom."	Billy	Mother
2	"Is everything okay at home?"	Billy	Mother
3	"Yes. Everything's fine."	Billy	Mother
4	"Where's Dad?"	Billy	Mother
5	"She's in the living room."	Billy	Mother
6	"How about Grandma and Grandpa?"	Billy	Mother
7	"They're in the yard."	Billy	Mother
8	"And where are you?"	Billy	Mother
9	"So everybody is fine?"	Billy	Mother
10	"Yes, Mom. We're all fine."	Billy	Mother

WHERE IS EVERYBODY?

They're	I'm	He's	She's

1 Where's Dad? _____He's_____ in the basement.

2 Where's Susie? _____ in the living room.

3 Where are Grandma and Grandpa? _____ in the yard.

4 Where are you, Billy? _____ in the kitchen.

CLOSE-UP

Where are you right now?

....

The video editor made a mistake! Put the following lines in the correct order.

____ Where's Dad?

____ I'm in the kitchen.

____ Yes, Mom. We're all fine.

____ She's in the living room.

____ Yes. Everything's fine.

____ How about Grandma and Grandpa?

__1__ Is everything okay at home?

____ So everybody's fine?

____ They're in the yard.

____ He's in the basement.

____ And Susie?

____ I'll see you soon, Billy.

____ And where are you?

WRITE THE SCRIPT!

You're calling home. Using the following as a guide, write the script, and then practice it with a friend.

A. Hi, ... This is

....

B. Oh hi, ...

A. Is everything okay at home?

B. Yes, everything's fine.

A. Where's ....?

B. .... in the ....

A. How about .... and

.... ?

B. .... in the ....

A. And where are you?

B. .... in the ....

A. So everybody's fine?

B. Yes, we're all fine.

A. That's good. I'll see you soon,

....

B. Bye.

SCRAMBLED SOUND TRACK

The sound track is all mixed up. Put the words in the correct order.

1. living room . She's the in   *She's in the living room.*

2. you ? are Where

3. in the car is . The garage

4. kitchen Mrs. Wong in . the is

5. and Paul are ? Where Tom

6. favorite What's room your ?

7. is Where Mrs. Ramirez ?

8. is Fred . in basement the

9. okay ? home Is at everything

WHAT'S MY LINE?

1. Where (is / are) Jane?

2. I (are / am) in the dining room.

3. Where (am / are) Carlos and Mike?

4. Where (are / is) your car?

5. Mr. Green (is / are) in the living room.

6. Where (is / are) you?

7. Where are Mr. (are / and) Mrs. Jones?

8. (We're / Where) in the kitchen.

SCRAMBLED WORDS

There are some problems with the sound track. Fix the scrambled words.

1. ngdini oorm   *dining room*

2. eaargg

3. rdya

4. dromobe

5. vilngi moro

6. sembatne

3:38 WHERE ARE YOU?

SON:	Where are you?
FATHER:	I'm in the kitchen.
DAUGHTER:	Where are you?
FATHER:	We're in the living room.
BOY 1:	Where are Mr. and Mrs. Jones?
BOY 2:	They're in the yard.

4:02 WHERE'S BOB?

WOMAN 1:	Where's Bob?
WOMAN 2:	He's in the living room.
WOMAN 3:	Where's Mary?
WOMAN 4:	She's in the bedroom.
SON:	Where's the car?
FATHER:	It's in the garage.

4:25 SBS-TV ON LOCATION

INTERVIEWER:	What's your favorite room at home?
PERSON 1:	The living room.
PERSON 2:	The dining room.
PERSON 3:	The kitchen.
PERSON 4:	My favorite room? My bedroom.
PERSON 5:	The basement.
INTERVIEWER:	The basement?
PERSON 5:	Yes. The basement.

4:40 IS EVERYTHING OKAY AT HOME?

BILLY:	Hello.
MOTHER:	Hi, Billy. This is Mom.
BILLY:	Oh hi, Mom.
MOTHER:	Is everything okay at home?
BILLY:	Yes. Everything's fine.
MOTHER:	Where's Dad?
BILLY:	He's in the basement.
MOTHER:	And Susie?
BILLY:	She's in the living room.
MOTHER:	How about Grandma and Grandpa?
BILLY:	They're in the yard.
MOTHER:	And where are you?
BILLY:	I'm in the kitchen.
MOTHER:	So everybody's fine?
BILLY:	Yes, Mom. We're all fine.
MOTHER:	That's good. I'll see you soon, Billy.
BILLY:	Okay.
MOTHER:	Bye.
BILLY:	Bye.

GRAMMAR

Subject Pronouns
To Be + Location

	am	I?
Where	is	he? she? it?
	are	we? you? they?

(I am)	I'm	
(He is) (She is) (It is)	He's She's It's	in the garage.
(We are) (You are) (They are)	We're You're They're	

FUNCTIONS

Asking for and Reporting Information

Is everything okay *at home?*
  Everything's fine.

Inquiring about Location

Where are *you?*
Where is *she?*
Where's *the car?*

Giving Location

I'm in the *kitchen.*
She's in the *bedroom.*
It's in the *garage.*

Greeting People

Hello.
Hi.

Leave Taking

Good-bye.
Bye.

I'll see you soon!

- **Places Around Town**
- **To Be: Subject Pronouns**

"Tell me the location and we'll have a conversation . . .

talking Side by Side."

PROGRAM LISTINGS

6:06 WHERE ARE THEY?
People are in several places around town.

7:00 ALL THE STUDENTS IN MY ENGLISH CLASS ARE ABSENT TODAY
Only one student is in English class today.

SBS-TV Backstage Bulletin Board

TO: Production Crew
Sets and props for this segment:

Restaurant
table
glass

Bank

Supermarket
groceries
shopping cart

Zoo
monkey

Library
book

Park
bench
book

Hospital
bed
blanket

Classroom
desks
blackboard
books
lights

TO: Cast Members
Key words in this segment:

restaurant
bank
supermarket
zoo
library
park

hospital
dentist
Social Security office
post office
absent
everybody

SOUND CHECK

I'm	We're	bank	restaurant
He's	You're	hospital	supermarket
She's	They're	library	zoo
It's		park	

1 A. Where's Albert?

B. ___He's___ in the ___restaurant___ .

2 A. Where's Carmen?

B. _____ in the _____ .

3 A. Where are Walter and Mary?

B. _____ in the _____ .

4 A. Where's the monkey?

B. _____ in the _____ .

5 A. Where are you?

B. _____ in the _____ .

6 A. Where are you?

B. _____ in the _____ .

7 A. Where am I?

B. _____ in the _____ .

Match the signs to the places below.

a. **Central Park**

b. **Boston City Hospital**

c. **NORTHWEST SAVINGS BANK**

d. **Children's Zoo**

e. **Los Angeles Public Library**

f. **Mario's Restaurant**

g. **$ Top Value $upermarket**

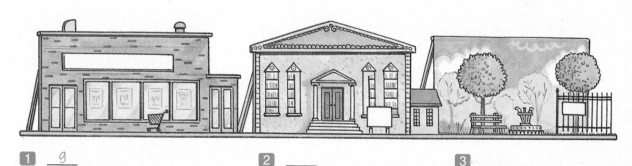

1. _g_   2. ___   3. ___

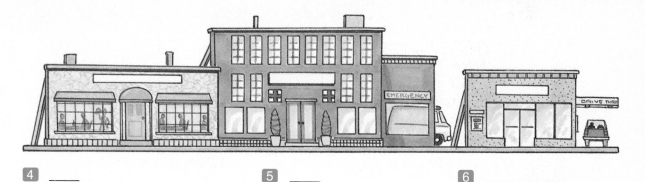

4. ___   5. ___   6. ___

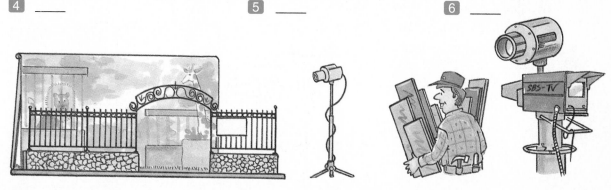

7. ___

7:00 ALL THE STUDENTS IN MY ENGLISH CLASS ARE ABSENT TODAY

1	George is absent today.	Yes	No
2	He's at the bank.	Yes	No
3	Maria is absent.	Yes	No
4	She's at the dentist.	Yes	No
5	Mr. and Mrs. Sato are absent.	Yes	No
6	They're in the hospital.	Yes	No
7	Our English teacher is in the restaurant.	Yes	No
8	Our English teacher is in the library.	Yes	No
9	All the students are in English class today.	Yes	No
10	All the students are at home.	Yes	No

CLOSE-UP

You're on Side by Side TV! All the students in your English class are absent today. Where are they?

....

....

....

WHAT'S MY LINE?

1 Where's (Tom) you and Bill ?
              they

2 Where are   the monkey
              Mr. Smith ?
              we

3 Where are   he
              Sue and Jane ?
              I

4 Where am   you
              I   ?
              Albert

5 Where's   Jack and Ted
              you   ?
              the car

6 Where's   George and I
              Rita   ?
              I

WHAT'S THE QUESTION?

| Where | Where's | am | are |

1 _____Where's_____ Peter?   He's in the supermarket.

2 _____ Paul and Mary?   They're in the park.

3 _____ the newspaper?   It's in the library.

4 _____ you and John?   We're in the living room.

5 _____ Helen?   She's in the hospital.

6 _____ you?   I'm in the kitchen.

7 _____ I?   You're in the basement.

8 _____ you?   ....

SCRAMBLED WORDS

There are some problems with the sound track. Fix the scrambled words.

1 The car is in the **agareg.**   *garage*

2 My favorite room is the **kcetnih.**   _____

3 Is Paul in the **blariyr?**   _____

4 Mr. and Mrs. Smith are in the **srepuktmera.**   _____

5 We're in the **stop coffie.**   _____

TV CROSSWORD

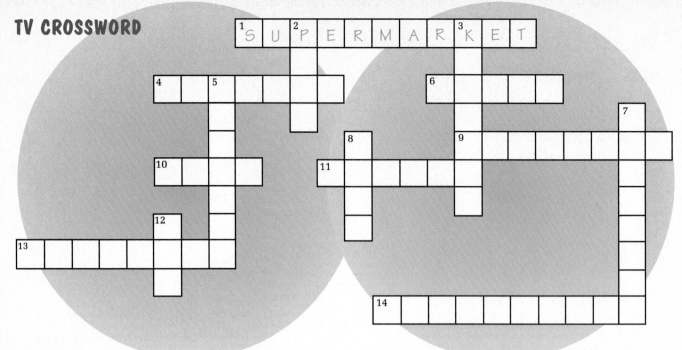

Across →

 1.  4.  6.  9.

 10.  11. 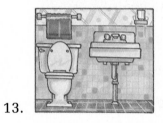 13.  14.

Down ↓

 2.  3.  5.

 7.  8.  12.

6:06 WHERE ARE THEY?

INTERVIEWER: Where's Albert?
PERSON 1: He's in the restaurant.

INTERVIEWER: Where's Carmen?
PERSON 2: She's in the bank.

INTERVIEWER: Where are Walter and Mary?
PERSON 3: They're in the supermarket.

INTERVIEWER: Where's the monkey?
PERSON 4: It's in the zoo.

INTERVIEWER: Where are you?
PERSON 5: I'm in the library.

INTERVIEWER: Where are you?
PERSON 6: We're in the park.

PATIENT: Where am I?
NURSE: You're in the hospital.
PATIENT: The...the hospital?
NURSE: Yes. You're in the hospital.
And you're okay.
PATIENT: I'm okay. I'm in the hospital,
and I'm okay.

7:00 ALL THE STUDENTS IN MY
ENGLISH CLASS ARE
ABSENT TODAY

STUDENT: All the students in my English
class are absent today. George is
absent. He's in the hospital. Maria
is absent. She's at the dentist. Mr.
and Mrs. Sato are absent. They're
at the Social Security office. Even
our English teacher is absent. He's
home in bed.

What a shame! Everybody in
my English class is absent today.
Everybody except me.

GRAMMAR

Subject Pronouns
To Be + Location

Where	am	I?
	is	he? she? it?
	are	we? you? they?

(I am)	I'm	
(He is) (She is) (It is)	He's She's It's	in the restaurant.
(We are) (You are) (They are)	We're You're They're	

FUNCTIONS

Inquiring about Location

Where are you?

Giving Location

He's in *the restaurant.*
They're at *the Social Security office.*

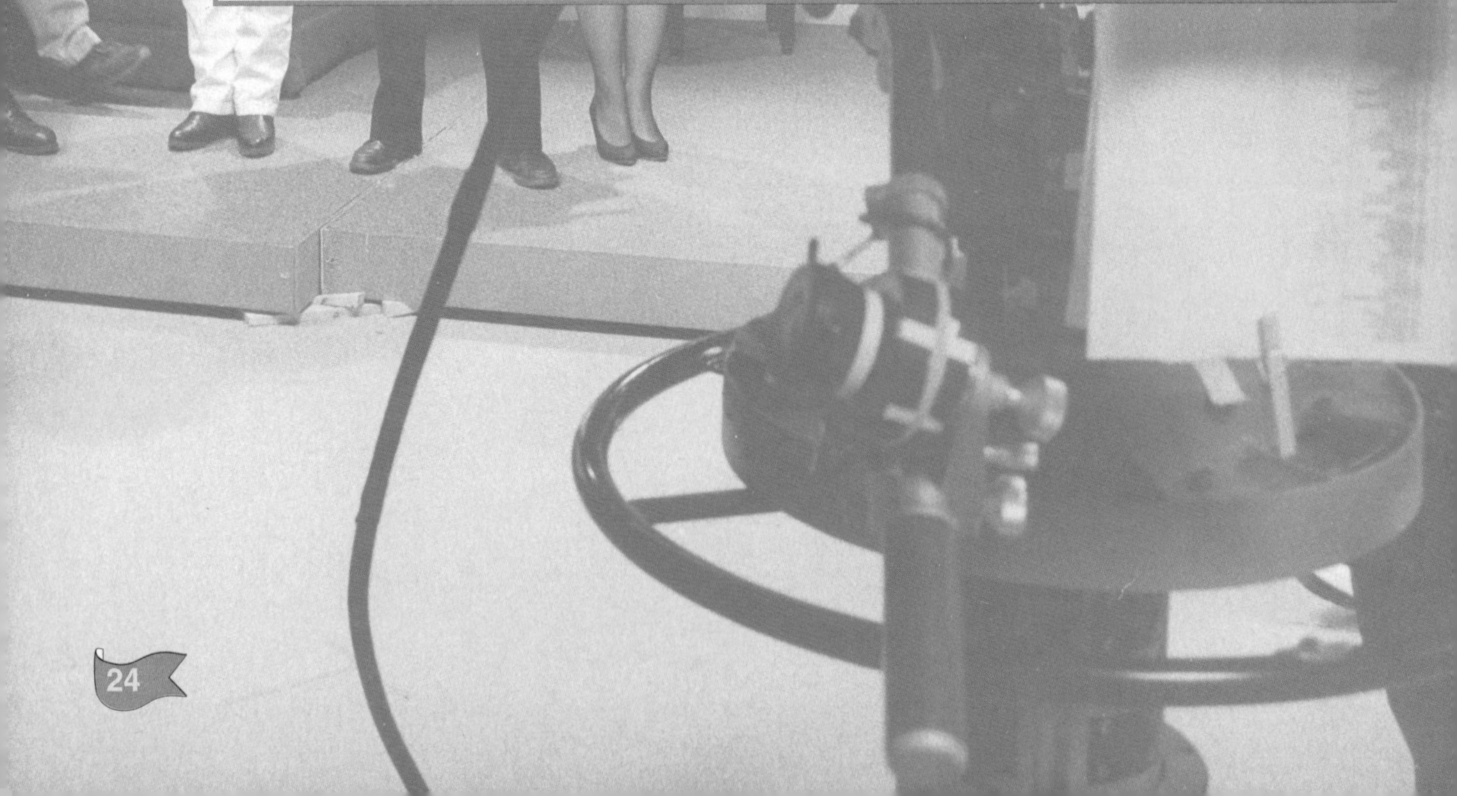

"Dancing, singing, cooking, eating, listening,

speaking, writing, reading . . . Side by Side."

PROGRAM LISTINGS

7:53 WHAT ARE YOU DOING?
People tell what they're doing.

8:38 WHERE'S CHARLIE?—
     GrammarRap
The GrammarRappers magically
appear in Charlie's kitchen.

9:09 WHAT'S EVERYBODY DOING?
People ask where others are.

9:37 WHERE'S BETTY?—GrammarRap
The GrammarRappers magically appear in
Betty's bedroom.

10:03 SBS-TV NEWS
Gary Carter, reporting live from Central Park,
interviews the Jones family. It's a beautiful day!

11:58 WHERE ARE MOM AND
     DAD?—GrammarRap
The GrammarRappers magically appear in
Mom and Dad's living room.

SBS-TV Backstage Bulletin Board

TO: Production Crew
Sets and props for this segment:

Bedroom
bed	book
dog	lamp
doll	
toys	

Dining Room
table
chairs
candles
glasses

Living Room
chair
sofa
lamps

Kitchen
table	cereal
chairs	orange juice
stove	sandwich
pots	glass of milk

Office
desks
telephones

Park
bench
newspaper
guitar

TO: Cast Members
Key words in this segment:

cooking
eating
listening
playing
reading
sleeping
studying

breakfast
lunch
dinner

music
guitar
newspaper

WHAT ARE YOU DOING?

SCRIPT CHECK

Help the actors prepare their lines.

| cooking | eating | reading | sleeping | studying | watching TV |

1 ___cooking___   **2** _____   **3** _____

4 _____   **5** _____   **6** _____

WHAT'S HAPPENING?

1 (a.) I'm reading.

   b. I'm eating.

2 a. We're reading.

   b. We're cooking.

3 a. They're studying English.

   b. They're sleeping.

4 a. He's reading.

   b. He's eating.

5 a. She's watching TV.

   b. She's reading.

6 a. It's eating.

   b. It's sleeping

CLOSE-UP

You're on Side by Side TV! Tell about yourself.

Where are you?   I'm ....

What are you doing?   I'm ....

FINISH THE RAP!

What's	lunch	He's	Eating	kitchen	Where's
		Who's	doing	in	

_____ _Where's_ [1] Charlie?

_____ [2] in the kitchen.

_____ [3] he doing?

_____ [4] lunch.

Charlie's in the kitchen eating

_____ [5].

Charlie's in the _____ [6]
eating lunch.

_____ [7] in the kitchen?

Charlie's _____ [8] the kitchen.

What's he _____ [9]?

Eating lunch.

WRITE YOUR OWN RAP!

Write a rap about Dave, and then practice it with a friend.

Dave's in the dining room... drinking tea.

Where's _____?

_____ in the _____.

What's he doing?

_____ing _____.

_____'s in the _____

_____ing _____.

_____'s in the _____ _____ing _____.

Who's in the _____?

Dave's in the _____.

What's he doing?

_____ing _____.

WHAT'S THE LINE?

Circle the lines you hear.

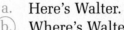

1

a.   Here's Walter.
b.   Where's Walter?
c.   She's in the kitchen.
d.   He's in the kitchen.
e.   What's he doing?
f.   What's she doing?
g.   She's eating breakfast.
h.   He's eating breakfast.

2

a.   Who's Betty?
b.   Where's Betty?
c.   She's in the yard.
d.   She's in the park.
e.   What are you doing?
f.   What's she doing?
g.   She's eating lunch.
h.   He's eating lunch.

3

a.   Here are Mr. and Mrs. Smith.
b.   Where are Mr. and Mrs. Smith?
c.   They're in the dining room.
d.   They're in the living room.
e.   What are they eating?
f.   What are they doing?
g.   They're eating lunch.
h.   They're eating dinner.

HELLO, EVERYBODY!

Where are you, Walter? What are you doing?

Hi, Betty! Where are you? What are you doing?

Mr. and Mrs. Smith, where are you? What are you doing?

1 __I'm__ in the _____ . **2** _____ in the _____ . **3** _____ in the _____ .

I'm _____ .   I'm _____ .   We're _____ .

> **FINISH THE RAP!**

she	bedroom	Who's	book	in	Where's
		reading	What's		

Where's ¹ Betty?

She's in the _____².

What's _____³ doing?

Reading a _____⁴.

Betty's in the bedroom _____⁵
a book.

Betty's _____⁶ the bedroom
reading a book.

_____⁷ in the bedroom?

Betty's in the bedroom.

_____⁸ she doing?

Reading a book.

> **WRITE YOUR OWN RAP!**

Write a rap about Lilly, and then practice it with a friend.

Lilly's in the living room…studying English.

Where's Lilly?

She's in the _____.

What's she doing?

_____ing _____.

Lilly's in the _____ _____ing

_____.

Lilly's in the _____ _____ing _____.

Who's in the _____?

_____'s in the _____.

What's she doing?

_____ing _____.

YES OR NO?

1	Gary Carter is in the park.	(Yes)	No
2	Bob and Jackie are in the park.	Yes	No
3	The Jones family is in the park.	Yes	No
4	It's a beautiful day in the park.	Yes	No
5	Mr. Jones is reading a book.	Yes	No
6	Mrs. Jones is listening to the radio.	Yes	No
7	Sally and Patty are singing.	Yes	No
8	Tommy Jones is playing cards.	Yes	No

ON CAMERA

You're a reporter for Side by Side TV! Report about another family: What are their names? Where are they? What are they doing?

Hello. This is .... reporting for

Side by Side TV News. I'm here in the ....,

and I'm talking with the .... family. It's a

beautiful day here in the ....

Mr. .... is ....ing.

Mrs. .... is ....ing. And

their children, ... and ....,

are ....ing. Yes. It's a beautiful day in the

....

Now practice the interview with some friends.

••••• SEGMENT 4

WHERE ARE MOM AND DAD?

FINISH THE RAP!

lunch	bedroom	in	What's	Watching	They're
	kitchen	are	Where's	He's	

Where ___*are*___ [1] Mom and Dad?

_____ [2] in the living room.

What _____ [3] they doing?

_____ [4] Channel Seven.

Betty's in the _____ [5].

Mom's in the living room.

Dad's _____ [6] the living room.

Charlie's in the _____ [7].

_____ [8] Charlie?

_____ [9] in the kitchen.

_____ [10] he doing?

Eating _____ [11].

YES OR NO?

1	Mom and Dad are in the dining room.	Yes	(No)
2	They're listening to the radio.	Yes	No
3	Mom and Dad are in the living room.	Yes	No
4	Betty's in the bedroom.	Yes	No
5	Charlie's in the kitchen.	Yes	No
6	Dad's in the kitchen.	Yes	No
7	Charlie is eating breakfast.	Yes	No

FINISH THE SCRIPT!

| cooking | listening | playing | reading | shining | studying |
| bedroom | kitchen | yard | dining room | living room |

It's a beautiful day today. The sun is

_____shining_____ [1]. Everybody in the Gomez

family is busy. Mr. Gomez is in the

_____ [2]. He's _____ [3]

the newspaper. Mrs. Gomez is in the

_____ [4]. She's _____ [5]

to music. Jenny Gomez is in the

_____ [6]. She's _____ [7]

English. Carla Gomez is in the

_____ [8]. She's _____ [9]

dinner. Edward Gomez is home, too. He's

_____ [10] the guitar in his

_____ [11]. It's a busy day in the

Gomez home today.

SCRAMBLED WORDS

There are some problems with the sound track. Fix the scrambled words.

1. Sam is **itenag** lunch in the dining room. _____eating_____

2. Helen and Howard are **estinlign** to music. _____

3. Maria is **tchwaing** TV in the living room. _____

4. Mrs. Ling is **diegran** the newspaper. _____

5. Jenny and Frank are **dncagin.** _____

6. Edward is **liynagp** the guitar. _____

7. What are Bob and Kathy **onidg?** _____

8. Tom is **kocigon** in the kitchen. _____

SCRAMBLED SOUND TRACK

There are some problems with the sound track. Fix the scrambled sentences.

1 doing | are | ? | What | they   _What are they doing?_

2 ? | doing | Sally | What's   _____

3 Luis | is | . | kitchen | eating | in | lunch | the   _____

4 watching | . | living room | TV | They 're | in | the   _____

5 book | a | reading | in | . | bedroom | the | I'm   _____

6 are | ? | Where | Peter | Tommy | and   _____

7 Antonio | cooking | dinner | . | the | in | is | kitchen   _____

8 in | is | bedroom | her | . | studying | Mariko   _____

WHAT'S MY LINE?

1 Steve ____ watching TV.   (is)   are

2 Nick and Judy ____ listening to music.   is   are

3 Elena ____ reading a book.   is   are

4 Mr. and Mrs. Molina ____ eating dinner.   is   are

5 Tony and Kim ____ dancing.   is   are

6 The sun ____ shining.   is   are

7 The birds ____ singing in the park.   is   are

8 Bob and Alice ____ cooking dinner.   is   are

CLOSE-UP

You're on Side by Side TV! Tell about yourself and your friends: Where is everybody? What's everybody doing?

....

....

....

....

....

SEGMENT 4 SCRIPT ●●●

7:53 WHAT ARE YOU DOING?

INTERVIEWER:	What are you doing?
PERSON 1:	I'm reading.
INTERVIEWER:	What are you doing?
PERSON 2:	We're cooking.
INTERVIEWER:	What are Mary and Fred doing?
PERSON 3:	They're studying English.
INTERVIEWER:	What's Tom doing?
PERSON 4:	He's eating.
INTERVIEWER:	What's Martha doing?
PERSON 5:	She's watching TV.
INTERVIEWER:	What's your dog doing?
PERSON 6:	It's sleeping.

9:09 WHAT'S EVERYBODY DOING?

FRIEND 1:	Where's Walter?
FRIEND 2:	He's in the kitchen.
FRIEND 1:	What's he doing?
FRIEND 2:	He's eating breakfast.
SECRETARY 1:	Where's Betty?
SECRETARY 2:	She's in the park.
SECRETARY 1:	What's she doing?
SECRETARY 2:	She's eating lunch.
BUTLER:	Where are Mr. and Mrs. Smith?
COOK:	They're in the dining room.
BUTLER:	What are they doing?
COOK:	They're eating dinner.

8:38 WHERE'S CHARLIE?—
GrammarRap

Where's Charlie?
  He's in the kitchen.
What's he doing?
  Eating lunch.
Charlie's in the kitchen eating lunch.
  Charlie's in the kitchen eating lunch.
Who's in the kitchen?
  Charlie's in the kitchen.
What's he doing?
  Eating lunch.

9:37 WHERE'S BETTY?—GrammarRap

Where's Betty?
  She's in the bedroom.
What's she doing?
  Reading a book.
Betty's in the bedroom reading a book.
  Betty's in the bedroom reading a book.
Who's in the bedroom?
  Betty's in the bedroom.
What's she doing?
  Reading a book.

SBS-TV NEWS

BOB ROGERS: It's a beautiful day today, Jackie.

JACKIE WILLIAMS: Yes, it is, Bob. It's a beautiful day. Before we go to the weather forecast, Side by Side TV News reporter Gary Carter is live in Central Park. Gary, are you there?

GARY CARTER: Yes, Jackie. I'm here in the park, and the sun is shining and the birds are singing. It's a beautiful day. With me in the park today is the Jones family. How do you like this beautiful weather, Mr. Jones?

MR. JONES: It's beautiful, Gary! It's a beautiful day!

GARY CARTER: What are you doing here in the park today?

MR. JONES: I'm reading the newspaper.

GARY CARTER: And how about you, Mrs. Jones? What are you doing today?

MRS. JONES: I'm listening to the radio.

GARY CARTER: How about you two? What are your names?

SALLY JONES: I'm Sally.

PATTY JONES: I'm Patty.

GARY CARTER: What are you kids doing today here in the park?

SALLY JONES: We're studying.

GARY CARTER: What's YOUR name?

TOMMY JONES: Tommy Jones.

GARY CARTER: And what are you doing on this beautiful day, Tommy?

TOMMY JONES: I'm playing the guitar.

GARY CARTER: Yes, Bob and Jackie. It's a beautiful day here in the park. The sun is shining and the birds are singing. Mr. Jones is reading the newspaper. Mrs. Jones is listening to the radio. Sally and Patty Jones are studying. And Tommy Jones is playing the guitar. Yes, it's a beautiful day. I'm Gary Carter, reporting live from Central Park for Side by Side TV News.

11:58 WHERE ARE MOM AND DAD?—GrammarRap

Where are Mom and Dad?
    They're in the living room.
What are they doing?
    Watching Channel Seven.
Betty's in the bedroom.
    Mom's in the living room.
Dad's in the living room.
    Charlie's in the kitchen.
Where's Charlie?
    He's in the kitchen.
What's he doing?
    Eating lunch.

GRAMMAR

Present Continuous Tense

What	am	I	doing?
	is	he she it	
	are	we you they	

(I am)	I'm	eating.
(He is) (She is) (It is)	He's She's It's	
(We are) (You are) (They are)	We're You're They're	

FUNCTIONS

Asking for and Reporting Information

What are you doing?
  I'm *reading*.

What's *Tom* doing?
  He's *eating*.

Who's *in the kitchen*?

Inquiring about Location

Where's *Charlie*?

Giving Location

He's in the *kitchen*.

SEGMENT 5

- **Daily Activities**
- **To Be: Short Answers**
- **Possessive Adjectives**

"They're so busy, yes they are. Busy fixing their old car and . . . talking Side by Side."

PROGRAM LISTINGS

12:43 I'M FIXING MY SINK
People call friends and family.

13:22 ARE YOU BUSY?
Lots of people are busy today.

14:12 I'LL CALL BACK
A telemarketer calls several people, but they're all busy.

SBS-TV Backstage Bulletin Board

TO: Production Crew
Sets and props for this segment:

Kitchen	Garage	Bedroom
sink	car	clothes
wrench	wrench	bed
telephone	telephone	telephone
mop	paint	
pail	paintbrushes	

Living Room	Dining Room	Bathroom
vacuum	table	sink
telephone	paper	shampoo
TV	pencils	

Office		Basement
telephones	flowers	bicycle
desks	paper	
photograph	water cooler	

TO: Cast Members
Key words in this segment:

my	busy	walking	dog
his	cleaning	washing	exercises
her	doing	apartment	hair
its	feeding	baby	homework
our	fixing	bicycle	room
your	painting	car	sink
their	talking	computer	windows

SOUND CHECK 1

Circle the words you hear.

1 (a.) fixing

b. washing

(c.) sink

2 a. fixing

b. car

c. garage

3 a. he's

b. cleaning

c. room

4 a. where

b. cleaning

c. apartment

5 a. children

b. homework

c. studying

SOUND CHECK 2

he's	I'm	she's	they're	we're	her	his
	my	our		their		

1 A. Hi! What are you doing?

B. ___I'm___ fixing ___my___ sink.

2 A. What's Bob doing?

B. _____ fixing _____ car.

3 A. What's Mary doing?

B. _____ cleaning _____ room.

4 A. What are you doing?

B. _____ cleaning _____ apartment.

5 A. What are your children doing?

B. _____ doing _____ homework.

13:22 ARE YOU BUSY?

THE WRONG CAPTIONS

*The caption writer made a mistake! Which captions are **wrong?***

1 Are you busy?

   (a.) Yes, I is.   b. Yes, I am.   (c.) Yes, I are.

2 Is Roger busy?

    a. Yes, we is.   b. Yes, he are.   c. Yes, he is.

3 Is Julie busy?

    a. Yes, she is.   b. Yes, she are.   c. Yes, they is.

4 Are you and your husband busy?

    a. Yes, we is.   b. Yes, we are.   c. Yes, I are.

5 Are Tom and Susan busy?

    a. Yes, she are.   b. Yes, they is.   c. Yes, they are.

SOUND CHECK

A. Are you busy?

B. Yes, I am. I'm (washing / watching)[1] my (hands / hair)[2].

A. Is Roger busy?

B. Yes, he (his / is)[3]. (He's / His)[4] cleaning (he's / his)[5] kitchen.

A. Is Julie busy?

B. Yes, (he / she)[6] is. (He's / She's)[7] fixing (her / the)[8] bicycle.

A. (Are / Our)[9] you busy?

B. Yes, we (are / our)[10]. We're doing (are / our)[11] exercises.

A. Are Tom and Susan busy?

B. Yes, they are. (Their / They're)[12] painting (their / they're)[13] garage.

WHAT'S EVERYBODY DOING?

1 (a.) Mr. Mifflin is fixing his computer.

   b. Mr. Mifflin is fixing his car.

2 a. Mrs. Mifflin is planting her garden.

   b. Mrs. Mifflin is painting her garage.

3 a. Mrs. Miller is cooking carrots.

   b. Mrs. Miller is cleaning the attic.

4 a. Mr. Miller is doing his homework.

   b. Mr. Miller is doing his exercises.

5 a. Mr. Ming is washing the windows.

   b. Mr. Ming is watching television.

6 a. Mrs. Ming is walking the dog.

   b. Mrs. Ming is washing the dog.

7 a. Lisa is talking to her cousin.

   b. Lisa is talking to a customer.

8 a. Dennis' wife is reading to the baby.

   b. Dennis' wife is feeding the baby.

EDITING MIX-UP 1

The video editor made a mistake! Put the conversation with the Miller family in the correct order.

____ Can I speak to Mrs. Miller, please?

____ Yes, it is.

 1 Hello.

____ Okay. I'll call back. Thank you.

____ Hello. Is this the Miller family?

____ I'm sorry. She's very busy right now.
She's cleaning the attic.

____ I'm afraid he's busy, too. He's doing his exercises.

____ Well then, can I possibly speak with Mr. Miller?

The video editor made another mistake! Put the conversation with the Ming family in the correct order.

_____ Okay. I'll call back. Thank you.

_____ Hello. Is this the home of the Ming family?

_____ Is Mr. Ming there?

_____ Is Mrs. Ming there?

_1_ Hello.

_____ Yes, she is, but she's busy, too. She's walking the dog.

_____ Yes, he is, but he's busy. He's washing the windows.

_____ Yes, it is.

MATCH THE LINES!

Match the sentences that have the same meaning.

_e_ **1** Hello.

_____ **2** I'm sorry. Mr. Jones is busy right now.

_____ **3** May I please speak to Mr. Jones?

_____ **4** Is this the home of the Jones family?

_____ **5** I'll call back.

a. Could I possibly speak with Mr. Jones?

b. Is this the Jones residence?

c. I'll call again later.

d. I'm afraid Mr. Jones is busy right now.

e. Hi.

WRITE THE SCRIPT!

The telemarketer is calling another family. Using the following as a guide, write the script, and practice it with a friend.

A. Hello.

B. Hello. Is this the .... family?

A. Yes, it is.

B. May I please speak to ....?

A. I'm sorry. He's busy right now. He's ....

B. Then may I speak to ....?

A. I'm very sorry. She's busy, too. She's ....

B. Okay. I'll call back. Thank you.

WRAP-UP

WRONG LINES

Cross out the mistakes.

1. I'm doing my ~~exercises~~ . homework / ~~sink~~

2. They're cleaning their ~~room~~ . garage / ~~homework~~

3. Mary is washing her ~~TV~~ . hair / ~~car~~

4. He's ~~reading~~ his dog. walking / feeding

5. She's painting her ~~apartment~~ . kitchen / ~~hair~~

6. He's ~~playing~~ his bicycle. fixing / ~~painting~~

7. We're cooking our ~~lunch~~ . dinner / ~~kitchen~~

8. Are you ~~washing~~ your sink? drinking / fixing

WHAT ARE THEY SAYING?

he	she	it	we	you	they	are	is

1. Are the children doing their homework?
   Yes, _____they are_____.

2. Are you eating dinner?
   Yes, _____.

3. Is your husband at home?
   Yes, _____.

4. Is Mrs. Smith there?
   Yes, _____.

5. Are you from Athens?
   Yes, _____.

6. Is your bicycle in the garage?
   Yes, _____.

TV CROSSWORD

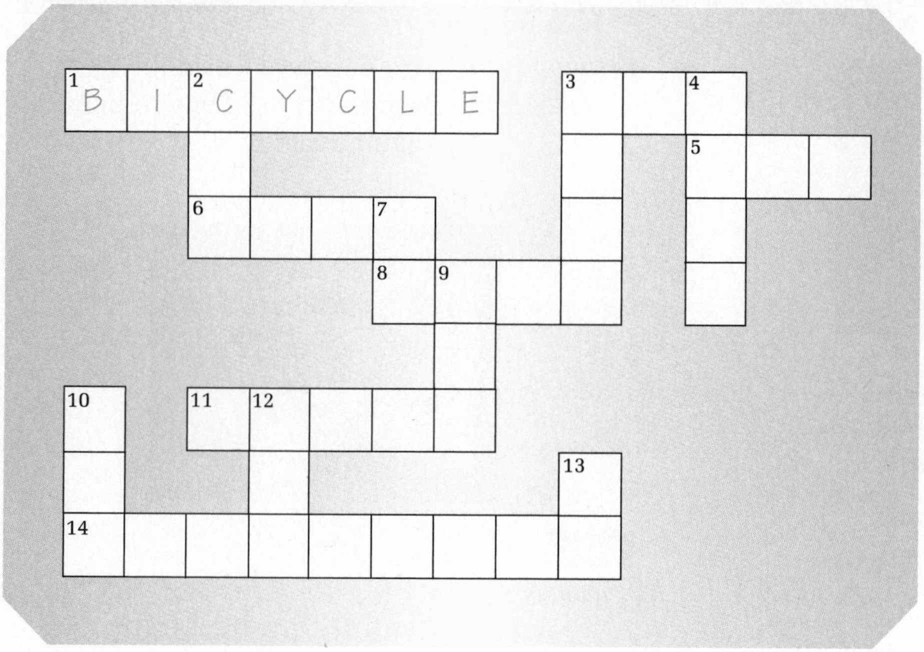

1 B I 2 C Y C L E
3
4
5
6
7
8
9
10
11
12
13
14

Across →

1. Lisa is in the yard. She's fixing her ____.
3. Tom is feeding ____ dog.
5. The dog is eating ____ dinner.
6. The children are cleaning their ____.
8. Are you doing ____ homework?
11. Mr. and Mrs. Lewis are painting ____ garage.
14. We're doing our ____.

Down ↓

2. Our ____ is in the garage.
3. I'm in the bathroom. I'm washing my____.
4. Mary is in the kitchen. She's fixing her ____.
7. I'm eating ____ lunch.
9. We're cleaning ____ apartment.
10. ____ you busy?
12. Mrs. Martin is fixing ____ computer.
13. ____ Julie busy?

SEGMENT 5 SCRIPT ●●●●●●●●●●●●●●●●●●●●●●●●●●●●●●

`12:43` I'M FIXING MY SINK

FRIEND 1: Hi! What are you doing?
FRIEND 2: I'm fixing my sink.

HUSBAND: What's Bob doing?
WIFE: He's fixing his car.

BOY: What's Mary doing?
FATHER: She's cleaning her room.

FRIEND 3: What are you doing?
FRIEND 4: We're cleaning our apartment.

FRIEND 5: What are your children doing?
FRIEND 6: They're doing their homework.

`13:22` ARE YOU BUSY?

INTERVIEWER: Are you busy?
PERSON 1: Yes, I am. I'm washing my hair.

INTERVIEWER: Is Roger busy?
PERSON 2: Yes, he is. He's cleaning his kitchen.

INTERVIEWER: Is Julie busy?
PERSON 3: Yes, she is. She's fixing her bicycle.

INTERVIEWER: Are you busy?
PERSON 4: Yes, we are. We're doing our exercises.

INTERVIEWER: Are Tom and Susan busy?
PERSON 5: Yes, they are. They're painting their garage.

`14:12` I'LL CALL BACK

PERSON 1: Hello.
DENNIS: Hello. Is this the Mifflin residence?
PERSON 1: Yes, it is.
DENNIS: May I please speak to Mr. Mifflin?
PERSON 1: I'm sorry. He's busy right now. He's fixing his computer.
DENNIS: Then may I speak to MRS. Mifflin?
PERSON 1: I'm sorry. She's busy, too. She's painting our garage.
DENNIS: Okay. I'll call back. Thank you.

(Dennis calls somebody else.)

PERSON 2: Hello.
DENNIS: Hello. Is this the Miller family?
PERSON 2: Yes, it is.
DENNIS: Can I speak to Mrs. Miller, please?
PERSON 2: I'm sorry. She's very busy right now. She's cleaning the attic.
DENNIS: Well then, can I possibly speak with MR. Miller?
PERSON 2: I'm afraid he's busy, too. He's doing his exercises.
DENNIS: Okay. I'll call back. Thank you.

(Dennis calls somebody else.)

PERSON 3: Hello.

DENNIS: Hello. Is this the home of the Ming family?

PERSON 3: Yes, it is.

DENNIS: Is Mr. Ming there?

PERSON 3: Yes, he is, but he's busy. He's washing the windows.

DENNIS: Is MRS. Ming there?

PERSON 3: Yes, she is, but she's busy, too. She's walking the dog.

DENNIS: Okay. I'll call back. Thank you.

(Dennis turns to talk with a co-worker.)

DENNIS: Lisa, I'm having some problems. I'm calling everybody, but they're all . . .

LISA: Dennis! I'm busy! I'm talking to a customer.

DENNIS: Sorry.

(Dennis calls his wife.)

WIFE: Hello.

DENNIS: Hi. It's me. I'm having a very strange day. Everybody's busy and . . .

WIFE: Dennis, honey? I'm sorry, but I'm very busy right now. I'm feeding the baby.

DENNIS: You're . . . you're busy?

WIFE: Yes. I'm very busy.

DENNIS: Oh . . . okay. I'll call you later.

(Dennis calls somebody else.)

DENNIS: Hello. I know you're very busy today. You ARE? Yes, I am too. I'm very busy today. It's a very busy day. Everybody's busy. I'm busy. You're busy. My wife is busy. My baby is busy. Busy! Busy! Busy! Busy! Busy! Busy!

GRAMMAR

To Be: Short Answers

Yes,	I	am.
	he she it	is.
	we you they	are.

Possessive Adjectives

I'm		my	
He's She's It's	cleaning	his her its	room.
We're You're They're		our your their	

FUNCTIONS

Greeting People

Hi!
Hello.

Asking for and Reporting Information

What are you doing?
  I'm *fixing my sink.*

Are you busy?
  Yes, I am. I'm *washing my hair.*

Is this the *Miller family?*
  Yes, it is.

Is *Mr. Ming* there?
  Yes, he is.

Leave Taking

I'll call you back.
I'll call back later.

Expressing Regret

I'm sorry.
Sorry.

I'm afraid *he's busy.*

Requesting

May I please speak to *Mr. Mifflin?*
Can I speak to *Mrs. Miller,* please?
Can I possibly speak with *Mr. Miller?*

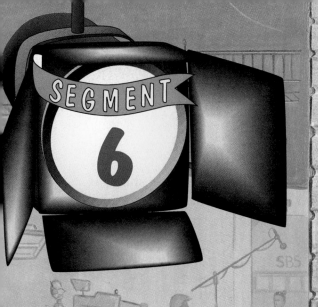

SEGMENT 6

- **Describing People, Places, and Things**
- **Adjectives**
- **Yes/No Questions**
- **Short Answers**

"Is he young or is he old? Is it hot or is it cold? . . .

talking Side by Side."

PROGRAM LISTINGS

SBS-TV Backstage Bulletin Board

TO: Production Crew
Sets and props for this segment:

Theater
director's chair

Laundromat
clothes
laundry baskets

Apartment
sofa
telephone
boxes

Living Room
table
chairs

Game Show
microphone
chairs
buzzer

TO: Cast Members
Key words in this segment:

tall	poor	easy
short	large	difficult
young	small	big
old	noisy	little
heavy/fat	quiet	nice
thin	married	interesting
handsome	single	happy
ugly	cheap	
rich	expensive	

17:35 TALL OR SHORT?

short	He's	Is	tall

A. Is Bob _____tall_____ [1] or short?

B. _____ [2] tall.

A. _____ [3] Bill tall or short?

B. He's _____ [4].

17:46 THE AUDITION

WHICH CAPTION?

Help the video editor match the caption with the adjective.

handsome	heavy	large	noisy	old	poor	quiet

rich	short	small	tall	thin	ugly	young

1 ___tall___

2 _____

3 _____

4 _____

5 _____

6 _____

7 _____

8 _____

9 _____

10 _____

11 _____

12 _____

13 _____

14 _____

19:21 TELL ME ABOUT...

YES, NO, OR MAYBE?

1	She's married.	Yes	(No)	Maybe
2	She's single.	Yes	No	Maybe
3	She's a student.	Yes	No	Maybe
4	His car is old.	Yes	No	Maybe
5	His car is large.	Yes	No	Maybe
6	His car is noisy.	Yes	No	Maybe
7	His neighbors are young.	Yes	No	Maybe
8	His neighbors are quiet.	Yes	No	Maybe
9	His neighbors are noisy.	Yes	No	Maybe

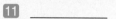

EVERYTHING'S FINE, MOM

SCENE CHECK

1 His name is . . .	Bill	Phil	(Neil)
2 She's his . . .	sister	mother	neighbor
3 His apartment isn't very . . .	cheap	nice	large
4 His apartment is very . . .	noisy	old	new
5 His neighbors aren't . . .	old	noisy	quiet
6 His new job is . . .	easy	difficult	quiet

WHOSE LINE?

1 "Hi, Neil. It's Mom."	Neil	(Mother)
2 "Dad and I are thinking about you."	Neil	Mother
3 "Everything's fine, Mom."	Neil	Mother
4 "Tell me about your apartment."	Neil	Mother
5 "Well, it isn't very large."	Neil	Mother
6 "No. It isn't new, Mom."	Neil	Mother
7 "And how about your neighbors?"	Neil	Mother
8 "It's a very quiet building."	Neil	Mother
9 "My new job is great!"	Neil	Mother
10 "Is the work easy or difficult?"	Neil	Mother
11 "I'm glad to hear that."	Neil	Mother
12 "Thanks for calling, Mom."	Neil	Mother

THE NEXT LINE

Circle the right response.

1 Hi, Mom. How are you?

(a.) We're fine.

b. Hello.

2 Tell me about your apartment. Is it large?

a. Everything's fine, Mom.

b. It isn't very large, but it isn't very small.

3 Is it new?

a. No. It isn't new, Mom.

b. That's good.

4 And how about your neighbors?

a. My neighbors?

b. We're fine.

5 Tell me about your new job.

a. My new job is great!

b. Very happy.

6 Speak to you soon.

a. Hello.

b. Good-bye.

A LETTER FROM NEIL

cheap	quiet	small	interesting
happy	large	noisy	difficult

Hi!

 I'm living in Chicago now. It's really a wonderful city. My apartment is nice. It isn't very ____large____ [1], but it isn't very _____ [2]. It's a really great apartment, and it isn't expensive. It's very _____ [3]. My neighbors are terrific. They aren't _____ [4]. In fact, the whole building is very _____ [5]. Also, my new job is great! The work is _____ [6], but very _____ [7]. I'm very _____ [8] here.

 How about you? Write back and tell me.

 Neil

Write a letter to Neil and tell about yourself.

GETTING TO KNOW THE PLAYERS

| Little | Long | Young | Rich |

The host of *What's That Word?* is

Rich _____1 _____2.

One contestant is Lisa _____3.

She's from _____4 Beach, California.

The other contestant is Larry _____5.

He's from _____6 Rock, Arkansas.

WHOSE LINE?

1	"And here's the world's favorite game show host: Rich Young."	(Announcer)	Rich Young
2	". . . and welcome to another edition of *What's That Word?*"	Announcer	Rich Young
3	"Let's say 'Hello' to Lisa Long."	Rich Young	Lisa Long
4	"Hi, everybody."	Rich Young	Lisa Long
5	"Tell us, where are you from?"	Rich Young	Lisa Long
6	"Let's say 'Hello' to Larry Little."	Rich Young	Larry Little
7	"Larry, welcome to *What's That Word?*"	Rich Young	Larry Little
8	"Hi, Rich. Hi, everybody."	Rich Young	Larry Little
9	"Are you ready, Lisa and Larry?"	Rich Young	Lisa and Larry
10	"Ready, Rich."	Rich Young	Larry Little

OPPOSITE ADJECTIVES

Lisa Long and Larry Little are going to guess opposite adjectives. First, YOU try to guess the adjectives. Then watch to see who guesses the words on the show.

single	heavy	expensive	quiet	old	little	ugly	loud

Who guesses the word?

1	young	_old_	Lisa	(Larry)
2	handsome	_____	Lisa	Larry
3	noisy	_____	Lisa	Larry
4	married	_____	Lisa	Larry
5	big	_____	Lisa	Larry
6	thin	_____	Lisa	Larry
7	cheap	_____	Lisa	Larry

THE BONUS ROUND!

Lisa is now playing the Bonus Round! First, YOU try to guess the adjectives. Then watch to see if Lisa guesses them correctly.

beautiful	tall	rich	thin	married	easy	large

Does Lisa guess the word?

1	short	_tall_	(Yes)	No
2	poor	_____	Yes	No
3	small	_____	Yes	No
4	ugly	_____	Yes	No
5	fat	_____	Yes	No
6	difficult	_____	Yes	No
7	single	_____	Yes	No

WHAT'S THAT ADJECTIVE?

1. My car isn't new. It's _____old_____.

2. My apartment isn't small. It's _____.

3. It isn't very cheap. It's _____.

4. I'm not married. I'm _____.

5. For me, mathematics isn't very easy.

   It's _____.

6. My mother isn't very tall. She's _____.

7. We aren't old. We're _____.

8. My neighbors aren't noisy. In fact, they're very _____.

TRUE OR FALSE?

1. A Rolls Royce is a very **cheap** car.   Yes, it is.   No, it isn't.

2. The Pacific Ocean is a **small** ocean.   Yes, it is.   No, it isn't.

3. The Pyramids are **old**.   Yes, they are.   No, they aren't.

4. The Amazon River is **short**.   Yes, it is.   No, it isn't.

5. The elephant is a **large** animal.   Yes, it is.   No, it isn't.

6. New York City is very **quiet**.   Yes, it is.   No, it isn't.

7. The Eiffel Tower is **tall**.   Yes, it is.   No, it isn't.

8. English is very **easy**.   ....

CLOSE-UP

You're on Side by Side TV! Tell about yourself, your neighbors, your family, your apartment, your work, and your school. Use the adjectives you learned in this segment and others you know.

....

....

....

....

`17:35` **TALL OR SHORT?**

INTERVIEWER: Is Bob tall or short?

BILL: He's tall.

INTERVIEWER: Is Bill tall or short?

BOB: He's short.

`17:46` **THE AUDITION**

DIRECTOR: All right. Who's next?

ACTOR: I am.

DIRECTOR: What's your name?

ACTOR: Lenny. Lenny Thomas.

DIRECTOR: Okay, Lenny Thomas. Let's see what you can do. I'm going to give you some adjectives, and you're going to act them out. Got it?

ACTOR: Yes. I understand.

DIRECTOR: Good. Let's begin. "Tall." "Short." "Young." "Old." "Heavy." "Thin." "Handsome." "Ugly." "Rich."

ACTOR: Hmm. Let me see.

DIRECTOR: "Poor." "Large." "Small." "Noisy." "Quiet."

ACTOR: "Shh!"

DIRECTOR: All right. That's all for today. Thank you, Benny.

ACTOR: Excuse me. It's Lenny. Lenny Thomas.

DIRECTOR: Okay, Lenny. Thank you. Next!

`19:21` **TELL ME ABOUT . . .**

MAN: Are you married?

WOMAN: No, I'm not. I'm single.

FRIEND 1: Tell me about your new car. Is it large?

FRIEND 2: No, it isn't. It's small.

FRIEND 3: Tell me about your new neighbors. Are they quiet?

FRIEND 4: No, they aren't. They're noisy.

`19:54` **EVERYTHING'S FINE, MOM**

SON: Hello.

MOTHER: Hi, Neil. It's Mom.

SON: Oh hi, Mom! How are you?

MOTHER: We're fine. Dad and I are thinking about you. We're wondering how you're doing there.

SON: Everything's fine, Mom. Everything's great.

MOTHER: Tell me about your apartment. Is it large?

SON: Well, it isn't very large, but it isn't very small. It's a really great apartment.

MOTHER: Is it new?

SON: No. It isn't new, Mom. In fact, it's very, very old. But it's really nice, and it's cheap.

MOTHER: That's good. And how about your neighbors?

SON:	My neighbors?
MOTHER:	Yes. Are they noisy or quiet?
SON:	They're very quiet. It's a very quiet building.
MOTHER:	That's nice. And tell me about your new job.
SON:	My new job is great!
MOTHER:	Is the work easy or difficult?
SON:	It's difficult, but very interesting.
MOTHER:	So you're happy there?
SON:	Very happy.
MOTHER:	I'm glad to hear that. Well, Neil, I'll say good-bye now. Just remember: Your Dad and I are thinking about you.
SON:	Thanks for calling, Mom. Speak to you soon.
MOTHER:	Good-bye.
SON:	Bye, Mom.

$1000

21:31 WHAT'S THAT WORD?

ANNOUNCER:	And now it's time to play the world's favorite game show, *What's That Word?* And here's the world's favorite game show host: Rich Young.
RICH YOUNG:	Thank you, ladies and gentlemen. Thank you. Thank you . . . and welcome to another edition of *What's That Word?* Let's meet our two contestants. First, let's say "Hello" to Lisa Long.
LISA:	Hi, everybody.
RICH YOUNG:	Lisa, welcome to *What's That Word?* Tell us, where are you from?

LISA:	I'm from Long Beach, California.
RICH YOUNG:	And now let's meet our other contestant. Let's say "Hello" to Larry Little.
LARRY:	Hi, Rich. Hi, everybody.
RICH YOUNG:	Larry, welcome to *What's That Word?* Tell us, where are you from?
LARRY:	I'm from Little Rock, Arkansas.
RICH YOUNG:	Okay, Lisa and Larry. It's time to play *What's That Word?* I'll say a word, and you press your buzzer and say the opposite. The contestant who says the most opposite words correctly wins the game . . . and gets one hundred dollars for each correct word. Are you ready, Lisa and Larry?
LISA:	Yes, I am.
LARRY:	Ready, Rich.
RICH YOUNG:	Then let's begin. "Young."
LARRY:	"Old."
RICH YOUNG:	That's correct, Larry. "Handsome."
LISA:	"Ugly."
RICH YOUNG:	That's right, Lisa. "Noisy."
LARRY:	"Loud."
RICH YOUNG:	Sorry, Larry.
LISA:	"Quiet."
RICH YOUNG:	Good, Lisa. "Married."
LARRY:	"Single."
RICH YOUNG:	That's correct. "Big."
LISA:	"Little."
RICH YOUNG:	Correct. "Thin."
LARRY:	"Heavy."
RICH YOUNG:	That's right. Our score is even. Lisa, you have three hundred dollars. And Larry, you have three hundred dollars. Are you ready for the last word?
LARRY:	Ready.
RICH YOUNG:	And the last word is "cheap."
LISA:	"Expensive."

RICH YOUNG: Congratulations, Lisa! You're the winner! Lisa, you'll continue with our bonus round in just a moment. Larry, let's listen to what we have for you.

ANNOUNCER: Larry, for playing *What's That Word?* you're going home with three hundred dollars. And to help you with those opposites, we're also giving you a copy of the Molinsky and Bliss portable *Pocket Dictionary.*

RICH YOUNG: Larry, we're sorry to have to say good-bye. Thanks for playing *What's That Word?*

(Larry walks off.)

RICH YOUNG: And now, Lisa, it's time for the bonus round. You have fifteen seconds. I'll give you a word, and you say the opposite word as quickly as you can. For each correct opposite, you get another one hundred dollars. Are you ready, Lisa?

LISA: Yes, I am.

RICH YOUNG: Then let's begin. "Short."

LISA: "Tall."

RICH YOUNG: "Poor."

LISA: "Rich."

RICH YOUNG: "Small."

LISA: "Large."

RICH YOUNG: "Ugly."

LISA: "Beautiful."

RICH YOUNG: "Fat."

LISA: Uh. . . uh. . . Go to the next one.

RICH YOUNG: "Difficult."

LISA: "Easy."

RICH YOUNG: "Single."

LISA: "Married."

RICH YOUNG: Lisa, listen to what you've won!

ANNOUNCER: Lisa, you have six correct opposites! That's another six hundred dollars. Your grand total today is one thousand dollars!

RICH YOUNG: Well, Lisa, you're our big winner today. How do you feel?

LISA: Happy. Very happy.

RICH YOUNG: Well, folks, that's all for today. Remember, whether you're young or old, rich or poor, tall or short, married or single, you're all welcome here on *What's That Word?* I'm your host, Rich Young, saying good-bye for now. See you next time.

ANNOUNCER: *What's That Word?* is a Side by Side Television Production.

GRAMMAR

To Be: Yes/No Questions

Am	I	
Is	he she it	tall?
Are	we you they	

To Be: Short Answers

Yes,	I	am.
	he she it	is.
	we you they	are.

No,	I'm	not.
	he she it	isn't.
	we you they	aren't.

FUNCTIONS

Asking for and Reporting Information

Is Bob *tall* or *short?*
  He's *tall.*

Tell me about *your new car.*

How about *your new job?*

Are you *happy* there?

What's *your name?*

Are *you married?*

How are *you?*

Describing

He's *tall.*

My *job* is *interesting.*

Greeting People

Hi, *Neil.*

Leave Taking

Speak to you soon.
Good-bye.
Bye.

Expressing Satisfaction

I'm glad to hear that.

My job is great.
Everything's fine/great.

Agreeing

That's correct.
That's right.
Correct.

Congratulating

Congratulations!

Identifying

It's *Mom.*

SEGMENT 7

"It's cloudy and it's raining, but we're really not complaining . . . 'cause we're Side by Side."

PROGRAM LISTINGS

SBS-TV Backstage Bulletin Board

TO: Production Crew
Sets and props for this segment:

Hall
  closet
  sunglasses
  shirt
  umbrella
  hat
  jacket
  scarf

Hotel/Bedroom
  telephone
  bed

TV Studio
  weather map

TO: Cast Members
Key words in this segment:

hot
warm
cool
cold
sunny
cloudy
raining
snowing

weather
vacation
having a good/
    terrible time
world
temperature
Celsius
Fahrenheit

 **HOW'S THE WEATHER TODAY?**

WEATHER CHECK

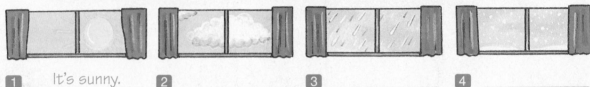

| It's | cloudy | cool | hot | raining | snowing | cold |
| | | sunny | warm | | | |

1. It's sunny.   2. _____   3. _____   4. _____

5. _____   6. _____   7. _____   8. _____

YES OR NO?

| Yes, it is. | No, it isn't. |

1. Is it cloudy?   2. Is it raining?   3. Is it warm?

No, it isn't.   _____   _____

4. Is it sunny?   5. Is it hot?   6. Is it snowing?

_____   _____   _____

Watch the scene and then practice it with a friend using this script.

ACTOR:

How's the weather today?

Oh, good!

Oh, okay.

It's  ?

Oh. It isn't raining.

Well, that's good.

It's ❄ ?!

Okay. I like snow.

It isn't?

It's 🌡 ?!

Okay. It's 🌡 .

Oh. It's 🌡 . Okay.

You know . . . it isn't very 🌡 .

It's 🌡 .

It's 🌡 ?!

First it's ☀ , then it's ☁ .

Then it's 🌧 . Now it's ❄ .

Then it's 🌡 . Then it's 🌡 .

Then it's 🌡 . Now it's 🌡 .

Yes?

Thank you.

VOICE:

It's ☀ .

Hmm. No. It's ☁ .

Oh. Wait a minute. It's 🌧 .

Wait. I'm sorry. It isn't 🌧 .

It's ❄ .

Yes.

Oh. Wait a minute. It isn't ❄

any more.

No. It's 🌡 .

Yes. It's very 🌡 .

Well, maybe it isn't 🌡 . It's 🌡 .

You know . . . it isn't very 🌡 .

It's 🌡 .

I'm sorry. I'm wrong. It isn't 🌡 .

It's 🌡 .

Yes. It's VERY 🌡 !

Excuse me.

Have a nice day.

YES OR NO?

1 Jim is in Miami.	(Yes)	No
2 Jim is on vacation.	Yes	No
3 Jack is calling from Miami.	Yes	No
4 It's sunny in Miami.	Yes	No
5 It's cold in Miami.	Yes	No
6 Jim is having a good time.	Yes	No
7 The weather in Miami is terrible.	Yes	No

THE NEXT LINE

Circle the right response.

1 I'm calling from Miami.

   (a.) From Miami?

   b. Where are you?

2 What are you doing in Miami?

   a. I'm calling from Miami.

   b. I'm on vacation.

3 Is it sunny?

   a. How's the weather?

   b. No, it isn't. It's raining.

4 Is it hot?

   a. No, it isn't. It's cold.

   b. Is it cold?

5 Are you having a good time?

   a. No, I'm not.

   b. It's raining.

6 The weather is TERRIBLE here.

   a. I'm happy to hear that.

   b. I'm sorry to hear that.

CLOSE-UP

You're on Side by Side TV! Tell the viewers: How's the weather today in YOUR city or town? Are you having a good time? What are you doing?

....

....

Dear Rita,

I'm writing to you from London, England. I'm on vacation with George and the children, and we're having a very good time.

The weather here is warm and sunny. How's the weather at home in Los Angeles?

We're staying at the Grand Hotel. Our rooms are very large and beautiful. Right now, I'm at the hotel watching TV and writing postcards. George and the children are at the London Zoo.

How are you? What are you doing?

See you soon,
Susan

YOUR POSTCARD TO A FRIEND

You're on vacation. Write your own postcard to a friend.

Dear ....,

    I'm writing to you from .... I'm on vacation with

........................, and we're having a .... time.

    The weather here is .... How's the weather at home in

.... ?

    We're staying at the .... Hotel. Our rooms are

.... Right now, I'm ....

....

How are you? What are you doing?

See you soon,

....

WORLD WEATHER UPDATE

WEATHER CHECK

1	*Mexico:*	cloudy	raining	(sunny)	snowing
2	*Canada:*	cloudy	raining	sunny	snowing
3	*England:*	cloudy	raining	sunny	snowing
4	*Japan:*	cloudy	raining	sunny	snowing
5	*Brazil:*	cold	cool	hot	warm
6	*Italy:*	cold	cool	hot	warm
7	*Australia:*	cold	cool	hot	warm
8	*Russia:*	cold	cool	hot	warm

WEATHER RECAP

Now it's time to recap today's weather.

1 _____It's sunny_____ in Mexico.   **5** _____ in Brazil.

2 _____ in Canada.   **6** _____ in Italy.

3 _____ in England.   **7** _____ in Australia.

4 _____ in Japan.   **8** _____ in Russia.

WEATHER CHALLENGE!

Watch the weather report for these 15 countries several times. How many answers can you give?

1	*Argentina:*	(hot)	cold	
2	*China:*	raining	snowing	
3	*Costa Rica:*	warm	cool	
4	*Germany:*	sunny	snowing	
5	*Greece:*	cool	warm	
6	*Hungary:*	cold	hot	
7	*Honduras:*	cold	cloudy	
8	*Korea:*	snowing	raining	

9	*Puerto Rico:*	cloudy	hot	
10	*Spain:*	warm	raining	
11	*Taiwan:*	sunny	cold	
12	*Turkey:*	cool	cloudy	
13	*Ukraine:*	snowing	raining	
14	*United States:*	cold	hot	
15	*Venezuela:*	raining	warm	

ON CAMERA

You're on TV! Tell about today's weather in different places around the world.

Hello, everybody. I'm .... with the World Weather Update.
(your name)

It's cold today in Canada. The temperature is 32°F/0°C. It's

....

....

....

....

And that's the World Weather Update. Have a nice day.

25:37 HOW'S THE WEATHER TODAY?

ACTOR: How's the weather today?

VOICE: It's sunny.

ACTOR: Oh, good!

VOICE: Hmm. No. It's cloudy.

ACTOR: Oh, okay.

VOICE: Oh. Wait a minute. It's raining.

ACTOR: It's raining?

VOICE: Wait. I'm sorry. It isn't raining.

ACTOR: Oh. It isn't raining. Well, that's good.

VOICE: It's snowing.

ACTOR: It's snowing?!

VOICE: Yes.

ACTOR: Okay. I like snow.

VOICE: Oh. Wait a minute. It isn't snowing any more.

ACTOR: It isn't?

VOICE: No. It's hot.

ACTOR: It's hot?!

VOICE: Yes. It's very hot.

ACTOR: Okay. It's hot.

VOICE: Well, maybe it isn't hot. It's warm.

ACTOR: Oh. It's warm. Okay.

VOICE: You know . . . it isn't very warm. It's cool.

ACTOR: You know . . . it isn't very warm. It's cool.

VOICE: I'm sorry. I'm wrong. It isn't cool. It's cold.

ACTOR: It's cold?!

VOICE: Yes. It's VERY cold!

ACTOR: First it's sunny, then it's cloudy. Then it's raining. Now it's snowing. Then it's hot. Then it's warm. Then it's cool. Now it's cold.

VOICE: Excuse me.

ACTOR: Yes?

VOICE: Have a nice day.

ACTOR: Thank you.

27:43 A LONG DISTANCE TELEPHONE CALL

JIM: Hi, Jack. This is Jim. I'm calling from Miami.

JACK: From Miami? What are you doing in Miami?

JIM: I'm on vacation.

JACK: How's the weather in Miami? Is it sunny?

JIM: No, it isn't. It's raining.

JACK: Is it hot?

JIM: No, it isn't. It's cold.

JACK: Are you having a good time?

JIM: No, I'm not. I'm having a TERRIBLE time. The weather is TERRIBLE here.

JACK: I'm sorry to hear that.

28:16 WORLD WEATHER UPDATE

ANNOUNCER: Now here's the World Weather Update from Side by Side TV News, with Side by Side meteorologist Maria Hernandez.

MARIA HERNANDEZ: Hello, everybody. It's a very interesting day today for weather around the world. It's sunny in Mexico. It's cloudy in Canada. Here in England, it's raining. And today in Japan, it's snowing. And look at that temperature in Brazil: a hundred degrees Fahrenheit, thirty-eight degrees Celsius! It's HOT!

It's warm today in Italy. The temperature there is seventy degrees Fahrenheit, twenty-one degrees Celsius. It isn't warm in Australia: fifty degrees Fahrenheit, ten degrees Celsius. It's cool! And it's cold today in Russia. The temperature there is thirty-two degrees Fahrenheit, zero degrees Celsius. Now here's the weather in OTHER parts of the world.

(Weather information.)

Argentina:	*hot*
China:	*raining*
Costa Rica:	*warm*
Germany:	*snowing*
Greece:	*cool*
Hungary:	*cold*
Honduras:	*cloudy*
Korea:	*snowing*
Puerto Rico:	*hot*
Spain:	*raining*
Taiwan:	*sunny*
Turkey:	*cool*
Ukraine:	*snowing*
United States:	*cold*
Venezuela:	*warm*

MARIA HERNANDEZ: And that's the World Weather Update from Side by Side TV News. I'm Maria Hernandez. Have a nice day.

GRAMMAR

To Be: Yes/No Questions

Is it sunny?

To Be: Short Answers

Yes, it is.

No, it isn't.

Adjectives

sunny	warm
cloudy	cool
hot	cold

FUNCTIONS

Asking for and Reporting Information

I'm calling from *Miami.*

What are you doing *in Miami?*

How's the weather *in Miami?*
  It's *raining.*

Is it *hot?*
  No, it isn't. It's *cold.*

It's *cold* today in *Russia.*

The temperature is *32°*
    Fahrenheit/*0°* Celsius.

Attracting Attention

Excuse me.

Wait a minute.

Apologizing

I'm sorry.

Correcting

It isn't *cool.* It's *cold.*

Greeting People

Hi, *Jack.* This is *Jim.*

Hello, *everybody.*

Expressing Agreement

Oh, okay.
Okay.

Expressing Dissatisfaction

I'm having a terrible time.

Expressing Satisfaction

Oh, good!
Well, that's good.

Expressing Surprise

It's snowing?!

Leave Taking

Have a nice day.

Sympathizing

I'm sorry to hear that.

SEGMENT 8

- **Family Members**
- **Prepositions of Location**
- **Present Continuous Tense Review**

"All the fathers and the mothers, all the sisters and the brothers . . . talking Side by Side."

PROGRAM LISTINGS

SBS-TV Backstage Bulletin Board

TO: Production Crew
Sets and props for this segment:

Office
photograph
desk
picture

Hallway
photographs
picture frames

Living Room
photo album
sofa
lamp

TO: Cast Members
Key words in this segment:

mother	grandmother	crying
father	grandfather	dancing
wife	grandparents	living
husband	aunt	looking
sister	uncle	playing
brother	cousin	smiling
daughter		working
son		

MY FAVORITE PHOTOGRAPHS

Help the cast rehearse important words in this segment.

Margaret Paul   Mary   Robert

Tina Bill   Michael   Jane   Richard

Timmy   Julie

husband	daughter	wife	son

Jane is Michael's _____ *wife* _____ 1.   Timmy is their _____ 3.

Michael is her _____ 2.   Julie is their _____ 4.

parents	brother	father	sister	mother

Bill is Michael's _____ 5.   Margaret and Paul are Michael's

Tina is his _____ 6.   _____ 7. Margaret is his

_____ 8, and Paul is his

_____ 9.

grandparents	cousin	aunt	grandmother	uncle	grandfather

Paul's sister Mary is Michael's   Margaret and Paul are Timmy and Julie's

_____ 10. Her husband   _____ 13. Margaret is

is Michael's _____ 11.   their _____ 14, and Paul

Their son Richard is Michael's   is their _____ 15.

_____ 12.

EDITING MIX-UP

The video editor made a mistake! Put the following conversation in the correct order.

_____ He's in Paris.

_____ His name is Paul.

__1__ Who is he?

_____ What's he doing?

_____ What's his name?

_____ This is a really nice photograph.

_____ Where is he?

_____ He's my father.

_____ He's standing in front of the Eiffel Tower.

_____ Thank you. It's my favorite photograph of my father.

WHAT'S THE RESPONSE?

__d__ **1** Who is she?   a. She's sitting in our living room.

_____ **2** What's her name?   b. It's my favorite photograph of my mother.

_____ **3** Where is she?   c. Her name is Margaret.

_____ **4** What's she doing?   d. She's my mother.

_____ **5** This is a really nice photograph.   e. She's knitting.

ALL IN THE FAMILY

1 My mother's sister is my _aunt_.

2 My mother's brother is my _____.

3 My father's mother is my _____.

4 My father's father is my _____.

5 My parents' other son is my _____.

6 Their daughter is my _____.

7 I'm married to Sally. She's my _____,

and I'm her _____.

8 We have a boy and a girl. They're our

_____ and our _____.

aunt	grandfather	sister	eating	dinner
brother	grandmother	son	playing	living room
cousin	husband	uncle	singing	restaurant
daughter	mother	wife	sleeping	yard
father			watching	They're

A. Who is she?

B. She's my _____mother_____ 1.

A. Oh, look! She's _____ 2 the piano.

B. Yes. She's _____ 3 the piano in

our _____ 4.

A. And is he your _____ 5?

B. Yes, he is.

A. What's he doing?

B. He's _____ 6 _____ 7

at his favorite _____ 8.

A. And who are they?

B. She's my _____ 9, Katie, and

he's my _____ 10, Matthew.

They're _____ 11

baseball in our _____ 12.

A. And who are they?

B. My _____ 13 and my _____ 14.

A. What are they doing? Are they _____ 15?

B. Yes. _____ 16 at my birthday party.

They're _____ 17 "Happy Birthday!"

A. Oh, what a nice photograph!

B. And this is Howard.

A. Your _____ 18?

B. Yes. He's my _____ 19.

A. Well, he isn't _____ 20 TV!

B. No, he isn't. He's _____ 21 in his favorite chair.

A. It's a great photograph.

A. And who is she?

B. She's my _____ 22's

_____ 23.

A. What's she _____ 24? Basketball?

B. No. She's _____ 25 volleyball.

B. And here's the rest of my family:

my _____ 26 and

my _____ 27,

my _____ 28

and _____ 29, and

my _____ 30. They're all

together for my _____ 31's birthday party.

A. What a nice family!

B. They're wonderful people.

A. Thanks for showing me your photo album.

B. My pleasure.

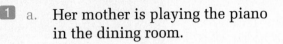
Do you remember what all the family members in the photographs are doing?

1
  a. Her mother is playing the piano in the dining room.

  b. Her brother is playing the piano in the living room.

  (c.) Her mother is playing the piano in the living room.

2
  a. Her father is eating lunch at his favorite restaurant.

  b. Her father is eating lunch in the dining room.

  c. Her father is eating dinner at his favorite restaurant.

3
  a. Her children are playing baseball in the garage.

  b. Her children are playing baseball in the park.

  c. Her children are playing baseball in the yard.

4
  a. Her sister and brother are singing.

  b. Her daughter and son are singing.

  c. Her friends are singing.

5
  a. Her husband is eating dinner.

  b. Her husband is sleeping.

  c. Her husband is watching TV.

6
  a. Her brother's wife is playing basketball.

  b. Her brother's wife is playing volleyball.

  c. Her brother's wife is playing baseball.

7
  a. The family is together for her grandmother's birthday party.

  b. The family is together for her cousin's birthday party.

  c. The family is together for her grandfather's birthday party.

EDITING MIX-UP

The video editor made a mistake! Put each pair of lines in the correct order.

1
  __2__ She's my mother.

  __1__ Who is she?

2
  ____ Yes. She's playing the piano in our living room.

  ____ Oh, look! She's playing the piano.

3
  ____ And is he your father?

  ____ Yes, he is.

4
  ____ And who are they?

  ____ My sister and my brother.

5
  ____ No, he isn't. He's sleeping in his favorite chair.

  ____ Well, he isn't watching TV!

6
  ____ My pleasure.

  ____ Thanks for showing me your photo album.

FINISH THE SONG!

smiling	living	having	looking	crying
	working	hanging	dancing	

I'm looking at the photographs.

They're hanging in the hall.

I'm _____smiling_____ [1] at the memories,

Looking at the pictures on the wall.

My son Robert's married now.

I'm _____ [2] in LA. (Hi, Dad!)

My daughter's _____ [3] in Detroit.

I'm very far away. (I love you, Dad!)

I'm _____ [4] at the photographs.

They're _____ [5] in the hall.

I'm smiling at the memories,

Looking at the pictures on the wall.

My mom and dad are _____ [6].

It's a very special day.

(We're _____ [7] a good time!)

My little sister's _____ [8].

It's my brother's wedding day.

(I'm so happy!)

I'm _____ [9] at the photographs.

They're _____ [10] in the hall.

I'm _____ [11] at the memories,

_____ [12] at the pictures on the wall.

I'm smiling at the memories,

Looking at the pictures on the wall.

1	His son Robert is single.	True	(False)
2	Robert is living in Detroit.	True	False
3	His daughter isn't living very near.	True	False
4	His daughter is dancing in the photograph.	True	False
5	His mom and dad are sad.	True	False
6	His mom and dad are smiling.	True	False
7	His little sister is dancing in the photograph.	True	False
8	It's her brother's wedding day.	True	False
9	The photographs are hanging in the living room.	True	False

RHYME TIME

What's the rhyming word from the song?

1 mad   _____Dad_____

2 tall   _____

3 car   _____

4 fun   _____

5 hat   _____

6 cooking   _____

7 day   _____

8 Mr.   _____

9 dry   _____

10 mother   _____

I'M ALL MIXED UP!

Unscramble the following lines from the song.

1 good We're time ! having a   _____We're having a good time!_____

2 the . looking at I'm photographs   _____

3 Detroit in daughter's My . working   _____

4 sister's little . My crying   _____

5 day special . It's very a   _____

6 away . very I'm far   _____

7 brother's . It's my day wedding   _____

8 I you Dad , love .   _____

9 son My Robert's . now married   _____

10 so ! I'm happy   _____

SCRAMBLED WORDS

There are some problems with the sound track. Fix the following family words.

1. Maria is my little **rsites.**   *sister*

2. My **dgrmtheoran** is having dinner. _____

3. What's your **augdterh** doing? _____

4. Where's your **terbhor?** _____

5. Peter's **osn** is in Rome. _____

6. Sara's **atnu** is watching TV. _____

7. Jenny's **elunc** is in the yard. _____

8. My **sinuco** is in the bank. _____

CLOSE-UP

You're on Side by Side TV! Draw your "family tree" and then tell about the people in your family.

30:08 MY FAVORITE PHOTOGRAPHS

CO-WORKER 1: Who is he?

CO-WORKER 2: He's my father.

CO-WORKER 1: What's his name?

CO-WORKER 2: His name is Paul.

CO-WORKER 1: Where is he?

CO-WORKER 2: He's in Paris.

CO-WORKER 1: What's he doing?

CO-WORKER 2: He's standing in front of the Eiffel Tower.

CO-WORKER 1: This is a really nice photograph.

CO-WORKER 2: Thank you. It's my favorite photograph of my father.

30:33 FAMILY PHOTOS

FRIEND 1: Who is she?

FRIEND 2: She's my mother.

FRIEND 1: Oh, look! She's playing the piano.

FRIEND 2: Yes. She's playing the piano in our living room.

FRIEND 1: And is he your father?

FRIEND 2: Yes, he is.

FRIEND 1: What's he doing?

FRIEND 2: He's eating dinner at his favorite restaurant.

FRIEND 1: And who are they?

FRIEND 2: She's my daughter, Katie, and he's my son, Matthew. They're playing baseball in our yard.

FRIEND 1: And who are they?

FRIEND 2: My sister and my brother.

FRIEND 1: What are they doing? Are they singing?

FRIEND 2: Yes. They're at my birthday party. They're singing "Happy Birthday!"

FRIEND 1: Oh, what a nice photograph!

FRIEND 2: And this is Howard.

FRIEND 1: Your husband?

FRIEND 2: Yes. He's my husband.

FRIEND 1: Well, he isn't watching TV!

FRIEND 2: No, he isn't. He's sleeping in his favorite chair.

FRIEND 1: It's a great photograph. And who is she?

FRIEND 2: She's my brother's wife.

FRIEND 1: What's she playing? Basketball?

FRIEND 2: No. She's playing volleyball. And here's the rest of my family: my grandmother and my grandfather, my aunt and uncle, and my cousin. They're all together for my grandmother's birthday party.

FRIEND 1: What a nice family!

FRIEND 2: They're wonderful people.

FRIEND 1: Thanks for showing me your photo album.

FRIEND 2: My pleasure.

32:16 PICTURES ON THE WALL—
Music Video

I'm looking at the photographs.
They're hanging in the hall.
I'm smiling at the memories,
Looking at the pictures on the wall.

My son Robert's married now.
  I'm living in L.A. (Hi, Dad!)
My daughter's working in Detroit.
  I'm very far away. (I love you, Dad!)

I'm looking at the photographs.
They're hanging in the hall.
I'm smiling at the memories,
Looking at the pictures on the wall.

My mom and dad are dancing.
  It's a very special day. (We're having
  a good time!)
My little sister's crying.
  It's my brother's wedding day.
  (I'm so happy!)

I'm looking at the photographs.
They're hanging in the hall.
I'm smiling at the memories,
Looking at the pictures on the wall.
I'm smiling at the memories,
Looking at the pictures on the wall.

GRAMMAR

To Be

Who is	he? she?
Who are	they?

He's my father. She's my mother.
They're my aunt and uncle.

Present Continuous Tense

What's	he she	doing?
What are	they	doing?

He's She's	studying.
They're	watching TV.

Prepositions of Location

She's **in** her bedroom.
He's **at** his favorite restaurant.
The pictures are hanging **on** the wall.

FUNCTIONS

Asking for and Reporting Information

Who is he?
  He's *my father.*
What's his name?
  His name is *Paul.*
What's he doing?
  He's *standing in front of the Eiffel Tower.*

Inquiring about Location

Where is *he?*

Giving Location

He's in *Paris.*

Complimenting

What a nice *photograph!*

Expressing Gratitude

Thank you.
Thanks for *showing me your photo album.*

Accepting Thanks

My pleasure.

- **Describing Location**
- **Places Around Town**
- **Prepositions of Location**
- **There Is/There Are**

" Around the corner from the pool, between the restaurant and the school . . . we're walking Side by Side."

PROGRAM LISTINGS

SBS-TV Backstage Bulletin Board

SBS-TV

TO: Production Crew
Sets and props for this segment:

Car
   steering wheel

Sidewalk
   laundry basket

Apartment Building Entrance
   mailboxes
   letter

TO: Cast Members
Key words in this segment:

bakery	across from
bank	around the corner from
hospital	between
laundromat	next to
library	neighborhood
movie theater	nearby
park	Excuse me.
police station	You're welcome.
post office	I'm new here.
restaurant	
school	
supermarket	

34:19 WHERE'S THE RESTAURANT?

SOUND CHECK

| across from | between | around the corner from | next to |

1 A. Where's the restaurant?

B. It's _____next to_____ the

bank.

2 A. Where's the supermarket?

B. It's _____ the

movie theater.

3 A. Where's the school?

B. It's _____ the

library and the park.

4 A. Where's the post office?

B. It's _____

the hospital.

ON CAMERA

Continue the scene. Ask and answer questions about places in YOUR neighborhood.

1 Where's .... ? It's ....

2 Where's .... ? It's ....

3 Where's .... ? It's ....

4 Where's .... ? It's ....

82

••••• SEGMENT 9

PICTURE THIS

Check (√) the correct picture.

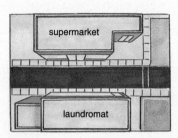 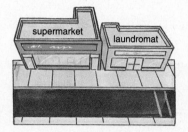 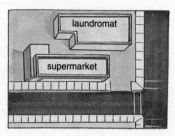

a. ____   b. ____   c. ____

SOUND CHECK

there	there's	next to
in	on	neighborhood

A. Excuse me. Is ____*there*____¹ a laundromat
____² this ____³?

B. Yes. ____⁴ a laundromat
____⁵ Main Street, ____⁶
the supermarket.

A. Thanks very much.

B. You're welcome.

CLOSE-UP

You're on Side by Side TV! Tell about YOUR neighborhood.

A. Is there a .... in your neighborhood?

B. Yes, there's a .... (next to/across from/around the corner
from/between) ....

A. Is there a .... in your neighborhood?

B. No, there isn't.

I'M NEW HERE

COMPLETE THE NEIGHBORHOOD MAP

Write the names on the correct buildings.

| apartment building | school | library | post office | movie theater |

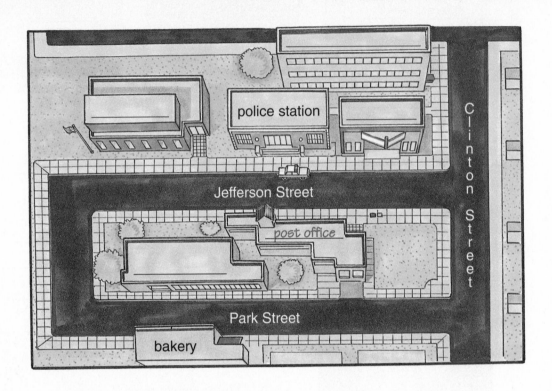

police station

Clinton Street

Jefferson Street

post office

Park Street

bakery

 DESCRIBE THE NEIGHBORHOOD

Using the map, write about the neighborhood.

across from	bakery
around the corner from	movie theater
between	police station
next to	post office
	school

1 The apartment building is ___*around the corner from the movie theater*___.

2 The library is _____.

3 The movie theater is _____.

4 The police station is _____.

5 The post office is _____.

●●●●● SEGMENT 9

next to	across from	there	post office	a
on	around	There's	school	am
		Where	neighborhood	is

A. Excuse me. Can I ask you a question?

B. Sure.

A. I'm new here. Is there a post office nearby?

B. Yes, there is. There's a post office on Jefferson Street, ____*across from*____ [1] the police station.

A. I see . . . and where's the police station?

B. It's between the _____ [2] and the movie theater.

A. Oh! Is _____ [3] a movie theater in this neighborhood?

B. Yes. _____ [4] a WONDERFUL movie theater here.

A. And WHERE _____ [5] it?

B. It's _____ [6] the police station.

A. And the police station is across from the _____ [7]. Right?

B. Yes. That's right.

A. And it's _____ [8] Jefferson Street.

B. Uh-húh.

A. Hmm. Where's Jefferson Street?

B. You ARE new here!

A. Yes, I _____ [9].

B. Jefferson Street is _____ [10] the corner from this building.

A. Oh!

B. Tell you what: I'm on my way to the library. Just come with me.

A. Oh, thank you.

B. No problem.

A. So there's _____ [11] library in this _____ [12]?

B. Yes. _____ [13] a fantastic library in this neighborhood.

A. _____ [14] is it?

B. It's _____ [15] the new bakery.

A. Oh. _____ [16] there a bakery nearby?

WHAT'S MY LINE?

A. Excuse me. Is [they / (there)]¹ a bank in this neighborhood?

B. Yes, [there's / where's]² a bank on River Street, next [from / to]³ the post office.

_____

A. Is there [the / a]⁴ post office [nearby / neighborhood]⁵?

B. Yes. There's a post office [on / in]⁶ Main Street, around the corner [to / from]⁷ the police station.

WHAT ARE THEY SAYING?

Using the map of the neighborhood, complete the conversations.

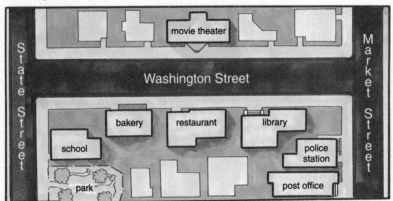

1 Excuse me. Where's the post office?

It's _____next to_____ the police station.

2 Is there a restaurant in this neighborhood?

Yes. There's a restaurant on Washington Street, _____ the bakery and the library.

3 Excuse me. Is there a bakery nearby?

Yes. It's _____ the restaurant, around the corner from the _____.

4 Where's the police station?

It's _____ the library.

5 _____?

It's on Washington Street, across from the restaurant.

6 Excuse me. Where's the school?

_____.

34:19   WHERE'S THE RESTAURANT?

HUSBAND 1:   Where's the restaurant?
WIFE 1:   It's next to the bank.

WIFE 2:   Where's the supermarket?
HUSBAND 2:   It's across from the movie theater.

WIFE 3:   Where's the school?
HUSBAND 3:   It's between the library and the park.

FRIEND 1:   Where's the post office?
FRIEND 2:   It's around the corner from the hospital.

34:56   IS THERE A LAUNDROMAT IN THIS NEIGHBORHOOD?

MAN: Excuse me. Is there a laundromat in this neighborhood?
WOMAN: Yes. There's a laundromat on Main Street, next to the supermarket.
MAN: Thanks very much.
WOMAN: You're welcome.

35:09   I'M NEW HERE

MAN: Excuse me. Can I ask you a question?
WOMAN: Sure.
MAN: I'm new here. Is there a post office nearby?
WOMAN: Yes, there is. There's a post office on Jefferson Street, across from the police station.
MAN: I see . . . and where's the police station?
WOMAN: It's between the school and the movie theater.
MAN: Oh! Is there a movie theater in this neighborhood?
WOMAN: Yes. There's a WONDERFUL movie theater here.
MAN: And WHERE is it?
WOMAN: It's next to the police station.
MAN: And the police station is across from the post office. Right?
WOMAN: Yes. That's right.
MAN: And it's on Jefferson Street.
WOMAN: Uh-húh.
MAN: Hmm. Where's Jefferson Street?
WOMAN: You ARE new here!
MAN: Yes, I am.
WOMAN: Jefferson Street is around the corner from this building.
MAN: Oh!
WOMAN: Tell you what: I'm on my way to the library. Just come with me.
MAN: Oh, thank you.
WOMAN: No problem.
MAN: So there's a library in this neighborhood?
WOMAN: Yes. There's a fantastic library in this neighborhood.
MAN: Where is it?
WOMAN: It's across from the new bakery.
MAN: Oh. Is there a bakery nearby?

GRAMMAR

Prepositions of Location

It's **next to** the bank.	It's **between** the library and the park.
It's **across from** the movie theater.	It's **around the corner from** the hospital.

There Is

Is there a laundromat in this neighborhood?	Yes, **there is.** No, **there isn't.**	**There's** a laundromat on Main Street.

FUNCTIONS

Inquiring about Location

Where's the *restaurant?*
Where is it?

Giving Location

It's next to *the bank.*
It's across from *the movie theater.*
It's between *the library* and *the park.*
It's around the corner from *the hospital.*

There's a *laundromat* on *Main Street,* next to *the supermarket.*

Asking for and Reporting Information

Is there a *laundromat* in *this neighborhood?*
Is there a *laundromat* nearby?
    Yes. There's a *laundromat* on *Main Street.*

Can I ask you a question?
    Sure.

Attracting Attention

Excuse me.

Checking Understanding

And it's on Jefferson Street?

Right?

Indicating Understanding

I see.

Oh!

Confirming

Yes. That's right.
Uh-húh.

Expressing Gratitude

Thank you.

Hesitating

Hmm.

- **Looking for an Apartment**
- **There Is/There Are**
- **Singular/Plural**

"There's a noisy elevator and an old refrigerator . . .

living Side by Side."

PROGRAM LISTINGS

SBS-TV Backstage Bulletin Board

TO: Production Crew

Sets and props for this segment:

Restaurant	*Realtor's Office*
tables	telephone
chairs	lamp
sandwiches	desk
	chairs

TO: Cast Members

Key words in this segment:

carpet	apartment	broken
closet	building	for rent
elevator	door	
refrigerator	sliding glass	
stove	door	
table	tub	
jacuzzi	lights	
washing	hall	
machine	wall	
window		

SCRIPT CHECK

Help the actors prepare their lines.

| stove | closet | elevator | jacuzzi | refrigerator | . table |
| washing machine | | window | | | |

1. _table_ 2. _____ 3. _____ 4. _____

5. _____ 6. _____ 7. _____ 8. _____

YES OR NO?

1. There's a nice apartment for rent.   (Yes)   No
2. There's a large window in the dining room.   Yes   No
3. There's a big closet in the bathroom.   Yes   No
4. The kitchen is big.   Yes   No
5. There isn't room for a table in the kitchen.   Yes   No
6. There's an elevator in the building.   Yes   No
7. There's a jacuzzi in the bedroom.   Yes   No

CLOSE-UP

You're on Side by Side TV! Tell about your house or apartment.

There's a/an .... in the living room.

There's a/an .... in the bedroom.

There's a/an .... in the kitchen.

There's a/an .... in the bathroom.

•••••• SEGMENT 10

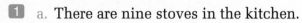

WHAT'S THE LINE?

Circle the lines you hear.

1
  a. There are nine stoves in the kitchen.

  (b.) There's a very nice stove in the kitchen.

  c. Oh, no!

  d. Oh, good!

2
  a. Is there a refrigerator in the kitchen?

  b. There isn't a refrigerator in the kitchen.

  c. Yes, there is.

  d. No, there isn't.

  e. Oh, I see.

WHAT'S THE LINE?

Circle the lines you hear.

  a. How many bathrooms are there in the apartment?

  b. There aren't two bedrooms in the apartment.

  c. Two bedrooms?

  d. How many closets are there in the apartment?

  e. There are three very large closets.

  f. Are there three washing machines in the building?

  g. But there's a laundromat around the corner.

WHICH ONE?

Which apartment are the people above talking about—Apartment A or Apartment B?

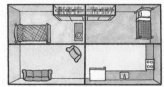

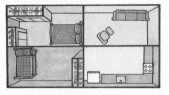

_____ **Apartment A**   _____ **Apartment B**

A new person is in the realtor's office. Write the script and then practice it with a friend.

A. Tell me, how many ....s are there in the apartment?

B. There are ....s in the apartment.

A. ....s?

B. Yes, that's right.

A. And tell me, are there many ....s in the apartment?

B. Yes, there are. There are .... very ....s.

A. Oh, good! And are there any ....s in the building?

B. No, there aren't. But there's a ....

   Are you interested in the apartment?

A. ....

THE NEXT LINE

1 Is there a stove in the kitchen?

   a. Yes, there are.

   (b.) Yes, there is.

2 Is there a refrigerator in the kitchen?

   a. No, there aren't.

   b. No, there isn't.

3 There's a very large living room.

   a. There is? Great!

   b. It is? Great!

4 How many bedrooms are there in the apartment?

   a. Yes, there are.

   b. There are two bedrooms.

5 Two bedrooms?

   a. Yes, that's right.

   b. Yes, it is.

6 Are there any washing machines in the building?

   a. No, there isn't.

   b. No, there aren't.

7 There's a jacuzzi in the bathroom.

   a. You're kidding!

   b. There isn't?

8 Are you interested in the apartment?

   a. Yes, there are.

   b. Yes, I am.

FINISH THE RAP!

| are | There | isn't | There's | aren't |

Tell me about the apartment on Elm Street.

It's nice, but it ___*isn't*___¹ very cheap.

_____² a brand new stove in the kitchen.

_____³ a beautiful carpet on the floor.

_____⁴ _____⁵ three large windows in the living room.

And the bedroom has a sliding glass door.

The bedroom has a sliding glass door?!

Yes! The bedroom has a sliding glass door.

Tell me about the apartment on Main Street.

It's cheap, but it _____⁶ very nice.

_____⁷ _____⁸ a tub in the bathroom.

_____⁹ _____¹⁰ any lights in the hall.

_____¹¹ a broken window in the dining room.

And _____¹² _____¹³ ten big holes in the wall!

_____¹⁴ are ten big holes in the wall?!

Yes! There _____¹⁵ ten big holes in the wall.

YES OR NO?

1	The apartment on Elm Street is expensive.	(Yes)	No
2	The apartment on Main Street isn't expensive.	Yes	No
3	The apartment on Elm Street is nice.	Yes	No
4	The apartment on Main Street is nice.	Yes	No
5	The stove in the kitchen on Elm Street isn't old.	Yes	No
6	There isn't a tub in the apartment on Main Street.	Yes	No
7	There are holes in the wall in the apartment on Elm Street.	Yes	No

SCRAMBLED WORDS

There are some problems with the sound track. Fix the scrambled words.

1 Is there a **oesvt** in the apartment?   _____

2 What a nice **otlesc!**   _____

3 There isn't an **oaeetvlr** in the building.   _____

4 What a nice **ettnapram!**   _____

5 Is there a **oaatrnulmd** nearby?   _____

6 There's room for a small **aetlb** in the dining room.   _____

7 Are there **hgiaswn eihsacmn** in the building?   _____

8 The **owwdni** in the bedroom is broken.   _____

9 A **azczuij?!** You're kidding!   _____

10 Is there a **aoiegfrrrret** in the apartment?   _____

WHAT'S THE LINE?

__j__ **1** I'm looking for an ____.   a. table

____ **2** Is there an apartment for ____?   b. broken

____ **3** There's a fireplace in the living ____.   c. door

____ **4** There are a lot of clothes in my ____.   d. kitchen

____ **5** Is there room in the kitchen for a ____?   e. corner

____ **6** There's a jacuzzi in the ____.   f. rent

____ **7** There's a large refrigerator in the ____.   g. room

____ **8** I'm sorry. The bedroom window is ____.   h. new

____ **9** The stove in the kitchen is brand ____.   i. interested

____ **10** The living room in this apartment has a sliding glass ____.   j. apartment

____ **11** There's a wonderful laundromat around the ____.   k. bathroom

____ **12** I'm very ____.   l. closet

SCRAMBLED SOUND TRACK

The sound track is all mixed up. Put the words in the correct order.

1 | a | very | . | living | There's | large | the | apartment | room | in |

There's a very large living room in the apartment.

2 | isn't | . | elevator | in | the | There | building | an |

3 | three | windows | . | large | the | dining | are | There | in | room |

4 | apartment | I'm | . | to | looking | rent | for | an |

5 | stove | in | apartment | ? | there | a | Is | the |

6 | like | Would | see | the | apartment | you | to | ? |

WHAT'S MY LINE?

1 There's / (There are) three bedrooms in the apartment.

2 Is / Are there a telephone in the apartment?

3 There's / There are a very large kitchen.

4 Are / Is there any stores near the apartment?

5 There are / There's a very new stove in the kitchen.

6 There isn't / aren't an elevator in the building.

7 How many rooms are / is there in the apartment?

8 Are / Is there an elevator in the building?

CLOSE-UP

You're on Side by Side TV! Tell about your home: How many rooms are there? What's in each room?

`36:30` AN APARTMENT FOR RENT

FRIEND 1: Are you still looking for an apartment?

FRIEND 2: Yes, I am. I'm still looking.

FRIEND 1: Well, there's a very nice apartment for rent in my building now.

FRIEND 2: Oh, really? Tell me about it.

FRIEND 1: There's a large window in the living room.

FRIEND 2: Hmm.

FRIEND 1: And there's a big closet in the bedroom.

FRIEND 2: Uh-húh?

FRIEND 1: And there's a great kitchen.

FRIEND 2: Is it large?

FRIEND 1: Yes. There's room for a nice big table.

FRIEND 2: Is there an elevator in the building?

FRIEND 1: No, there isn't. But get this: there's a jacuzzi in the bathroom!

FRIEND 2: You're kidding!

FRIEND 1: No, I'm serious! Are you interested in the apartment?

FRIEND 2: Yes. I'm very interested.

`37:20` IS THERE A STOVE IN THE KITCHEN?

RENTER 1: Is there a stove in the kitchen?

AGENT 1: Yes, there is. There's a very nice stove in the kitchen.

RENTER 1: Oh, good.

RENTER 2: Is there a refrigerator in the kitchen?

AGENT 2: No, there isn't.

RENTER 2: Oh, I see.

`37:37` HOW MANY BEDROOMS ARE THERE IN THE APARTMENT?

RENTER 3: Tell me, how many bedrooms are there in the apartment?

AGENT 3: There are two bedrooms in the apartment.

RENTER 3: Two bedrooms?

AGENT 3: Yes. That's right.

RENTER 3: And tell me, are there many closets in the apartment?

AGENT 3: Yes, there are. There are three very large closets.

RENTER 3: Oh, good! And are there washing machines in the building?

AGENT 3: No, there aren't. But there's a laundromat around the corner. Are you interested in the apartment?

RENTER 3: Yes, I am.

AGENT 3: Good.

`38:06` TELL ME ABOUT THE APARTMENT–GrammarRap

Tell me about the apartment on Elm Street.
  It's nice, but it isn't very cheap.
  There's a brand new stove in the kitchen.
  There's a beautiful carpet on the floor.
  There are three large windows in the living room.
  And the bedroom has a sliding glass door.
The bedroom has a sliding glass door?!
  Yes! The bedroom has a sliding glass door.

Tell me about the apartment on Main Street.
  It's cheap, but it isn't very nice.
  There isn't a tub in the bathroom.
  There aren't any lights in the hall.
  There's a broken window in the dining room.
  And there are ten big holes in the wall!
There are ten big holes in the wall?!
  Yes! There are ten big holes in the wall.

GRAMMAR

There Is/There Are

Is there a stove in the kitchen?
There's a very nice stove in the kitchen.
Yes, **there is.** No, **there isn't.**

Are there many closets in the bedroom?
There are three large closets.
Yes, **there are.** No, **there aren't.**

Singular

There's one bedroom in the apartment.

Is there	**a** washing machine? **an** elevator?

Plural

There are two bedrooms in the apartment.

There are two	washing machines. elevators.

FUNCTIONS

Asking for and Reporting Information

Is there a *stove* in the *kitchen?*
   Yes, there is.
   No, there isn't.

Are there any *closets* in the *bedroom?*
   Yes, there are.
   No, there aren't.

How many *bedrooms* are there in *the apartment?*
   There are *two bedrooms* in *the apartment.*

Tell me, _____?

Expressing Surprise

You're kidding!

Expressing Pleasure

Oh, good!

Checking Understanding

Two bedrooms?

Indicating Understanding

Uh-húh.

SEGMENT 11

- **Clothing**
- **Colors**
- **Singular/Plural**

"Buying coats and hats and skirts, pants and belts and ties and shirts . . . shopping Side by Side."

PROGRAM LISTINGS

SBS-TV Backstage Bulletin Board

TO: Production Crew
Sets and props for this segment:

Department Store
counter	dress	umbrella
shirt	hanger	watch
tie	coat	

Department Store/ Men's Jackets
hanger   a purple jacket

Department Store/ Women's Gloves
counter   a pair of green gloves

Clyde's Clothing Store
table	briefcase	belt
sweater	hat	necklace
umbrella		

TO: Cast Members
Key words in this segment:

belt	briefcase	black	pink	popular
dress	coat	blue	purple	a pair of
gloves	necklace	brown	red	need
hat	sweater	gold	silver	looking for
shirt	tie	gray	white	Can I help you?
umbrella	jacket	green	yellow	May I help you?
watch	color	orange	nice	

SHIRTS ARE OVER THERE

PROP DEPARTMENT

Help the production crew put together the props for this segment.

1 a s h i r t   **2** s h i r t s   **3** _ _ _ _ _   **4** _ _ _ _ _

5 _ _ _ _   **6** _ _ _ _   **7** _ _ _ _ _ _ _   **8** _ _ _ _ _ _ _ _

9 _ _ _ _ _ _   **10** _ _ _ _ _ _   **11** _ _ _ _ _   **12** _ _ _ _ _ _ _

WHOSE LINE?

1	"Excuse me. I'm looking for a shirt."	Salesperson	**Customer**
2	"Shirts are over there."	Salesperson	Customer
3	"Pardon me. I'm looking for a coat."	Salesperson	Customer
4	"Oh, hello. May I help you?"	Salesperson	Customer
5	"Dresses are right over there around the corner."	Salesperson	Customer
6	"Thanks very much."	Salesperson	Customer
7	"Can I help you?"	Salesperson	Customer
8	"You're welcome."	Salesperson	Customer

EDITING MIX-UP

The video editor made a mistake! Put the following conversation in the correct order.

_____ Yes, please. I'm looking for a dress.

__1__ Hello. How are you today?

_____ You're very welcome.

_____ Fine, thank you. May I help you?

_____ Thank you very much.·

_____ A dress? Dresses are right over there around the corner.

MATCH THE LINES!

Match the actors' lines that have the same meaning.

_e_ **1** Hello.	a.	Pardon me.
_____ **2** Excuse me.	b.	Thank you very much.
_____ **3** Thanks a lot.	c.	Can I help you?
_____ **4** May I help you?	d.	I need an umbrella.
_____ **5** I'm looking for an umbrella.	e.	Hi.

WRITE THE SCRIPT!

More customers are asking about things in the store. Write the script and then practice it with a friend.

A. Excuse me. I'm looking for a/an ....

B. .... s are over there.

A. Thank you.

B. You're welcome.

SBS-TV ON LOCATION

SOUND CHECK

What's your favorite color?

1 (a.) red   b. pink   **2** a. orange   b. red   **3** a. gold   b. I don't know.

4 a. orange   b. yellow   **5** a. red   b. green   **6** a. black   b. blue

7 a. pink   b. purple   **8** a. red   b. I don't know.   **9** a. black   b. blue

10 a. blue   b. silver   **11** a. yellow   b. pink   **12** a. gray   b. green

13 a. white   b. black   **14** a. silver   b. gold   **15** a. brown   b. black

CLOSE-UP

What's YOUR favorite color?

....

Interview two friends. What are their favorite colors?

1 Name: ....

Color: ....

2 Name: ....

Color: ....

42:51 I'M LOOKING FOR A JACKET

SOUND CHECK

MEN'S DEPARTMENT

A. May I help you?

B. Yes, please. I'm looking for a (jacket / jackets) [1].

A. Here's a nice jacket / jackets [2].

B. But this is a / an [3] purple jacket!

A. That's okay. Purple jacket / jackets [4] are very popular this year.

43:20 I'M LOOKING FOR A PAIR OF GLOVES

SOUND CHECK

A. Can I help you?

B. Yes, please. I'm looking for a pair of glove / gloves [1].

A. Here's a nice pair / pairs [2] of gloves.

B. But this / these [3] are green gloves!

A. That's okay. Green glove / gloves [4] are very popular this year.

Help the production crew find these props.

boots	earrings	pajamas	pants	shoes	socks

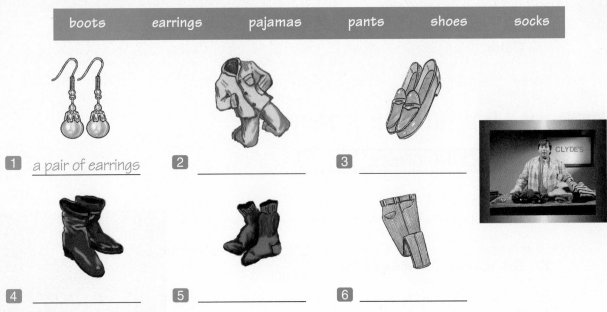

1 _a pair of earrings_

2 _____

3 _____

4 _____

5 _____

6 _____

43:37 CLYDE'S CLOTHING

Watch the commercial and then practice it using this script.

Hello. This is Clyde Clifford, owner of Clyde's Clothing. All of us here at Clyde's

Clothing are very proud of our store. At Clyde's, there's clothing for everyone. Are you

looking for _____ a hat _____[1]? Or maybe _____[2]?

How about _____[3]? What about _____[4]

for those rainy days? Are you looking for _____[5] for a special

person? What about a beautiful vinyl _____[6]? Well, come on over

to Clyde's, where there are lots of _____[7],

_____[8], _____[9], _____[10],

_____[11], and _____[12]. So take it from me,

Clyde Clifford. At Clyde's, there's clothing for everyone!

YOU'RE the owner of a clothing store! Using the script below, write and present a commercial for your store.

Hello. This is _____,
(your first and last name)

owner of _____'s Clothing. All of us here
(your first name)

at _____'s Clothing are very proud of our
(your first name)

store. At _____'s, there's clothing for everyone.

Are you looking for a/an _____? Or maybe a/an

_____? How about a/an _____?

What about a/an _____? Are you looking for a/an

_____? What about a beautiful _____?

Well, come on over to _____'s, where there are lots of

_____s, _____s, _____s,

_____s, _____s, and _____s!

So take it from me, _____. At _____'s,
(your first and last name)   (your first name)

there's clothing for everyone!

YOUR JINGLE: If your clothes are old,

                                  And you look a mess,

                                  Come to _____'s

                                  And dress for less.

                                  _____'s Clothing.

WHAT ARE THEY SAYING?

1 A. May I help you?

B. Yes, please. I'm looking for __a__

___blue___ ___jacket___ .

2 A. Can I help you?

B. Yes. I need _____ _____

_____ .

3 A. Here's _____ _____

_____ .

B. Thank you.

4 A. May I help you?

B. Yes, please. I'm looking for _____

_____ _____ .

5 A. Here's _____ nice _____

_____ .

B. Thanks very much.

6 A. Oh, hello. Can I help you?

B. Yes, please. I need _____

_____ _____ .

7 A. May I help you?

B. Yes, please. I need _____

_____ _____ .

8 A. Excuse me. I'm looking for _____

_____ _____ .

B. Here's _____ nice _____

_____ .

WHAT'S MY LINE?

1. It's raining. Here's (a / an) umbrella.

2. (Vinyl briefcases / Gold necklace) are very popular.

3. Are you looking for (a / an) nice sweater?

4. (Dress / Watches) are right over there.

5. I'm looking for a pair (from / of) gloves.

6. (These / This) bracelet (are / is) beautiful.

7. I need a pair of (shirts / pants).

8. I'm looking for an (orange / red) hat.

A CLOTHING STORE FLYER

Clyde's Clothing Store is having a sale. Write a flyer for the store.

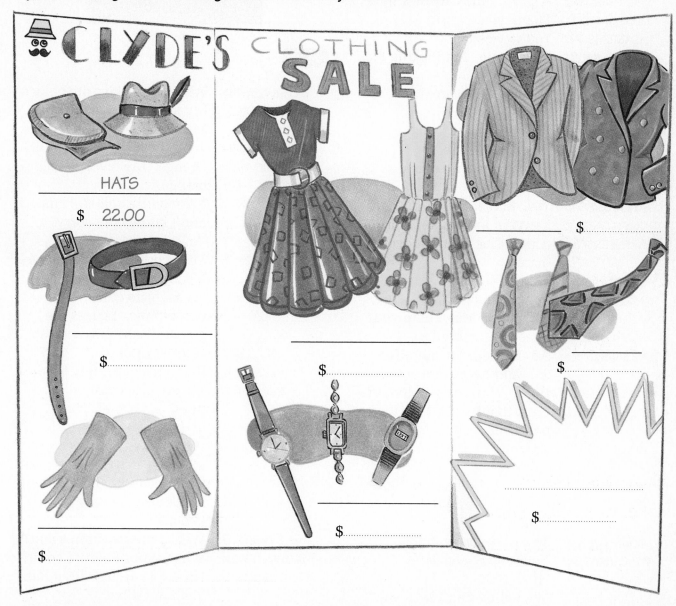

CLYDE'S CLOTHING SALE

HATS

$ 22.00

$ ....

$ ....

$ ....

$ ....

$ ....

$ ....

$ ....

39:10 SHIRTS ARE OVER THERE

CUSTOMER 1: Excuse me. I'm looking for a shirt.
SALESPERSON: A shirt? Shirts are over there.
CUSTOMER 1: Thank you.
SALESPERSON: You're welcome.
CUSTOMER 2: Pardon me. I'm looking for a coat.
SALESPERSON: A coat? Coats are over there.
CUSTOMER 2: Thank you.
SALESPERSON: You're welcome.
CUSTOMER 3: Excuse me. Are you a salesperson?
SALESPERSON: Yes. May I help you?
CUSTOMER 3: Yes. I'm looking for a tie.
SALESPERSON: A tie? Ties are right over there.
CUSTOMER 3: Thanks very much.
SALESPERSON: You're welcome.
Oh, hello. May I help you?
CUSTOMER 4: I need an umbrella.
SALESPERSON: An umbrella? Umbrellas are right over there.
CUSTOMER 5: Hello. How are you today?
SALESPERSON: Fine, thank you. May I help you?
CUSTOMER 5: Yes, please. I'm looking for a dress.
SALESPERSON: A dress? Dresses are right over there around the corner.
CUSTOMER 5: Thank you very much.
SALESPERSON: You're very welcome.
CUSTOMER 6: Hi!
SALESPERSON: Oh, hello. Can I help you?

CUSTOMER 6: Sure. I'm looking for a watch.
SALESPERSON: A watch? Watches are over there.
CUSTOMER 6: Thanks a lot.
SALESPERSON: You're welcome.

41:01 SBS-TV ON LOCATION

INTERVIEWER: What's your favorite color?
PERSON 1: My favorite color? Hmm. My favorite color is red.
INTERVIEWER: What's your favorite color?
PERSON 2: Orange. My favorite color is orange.
PERSON 3: My favorite color? Hmm. I don't know. I'm sorry. I don't know.
PERSON 4: My favorite color. Let me see. My favorite color is yellow.
PERSON 5: My favorite color? Gee. I don't know. I guess it's green. Yeah. My favorite color is green.
PERSON 6: Blue. It's definitely blue.
PERSON 7: My favorite color is . . . uhm . . . purple! That's it: purple.
PERSON 3: What's my favorite color? Maybe it's . . . hmm . . . no. Gee, I'm sorry. I don't know.
PERSON 8: Black. My favorite color is black.
PERSON 9: What's the question again?
INTERVIEWER: What's your favorite color?
PERSON 9: Hmm. I suppose it's silver. Yeah, it's silver.

PERSON 10: Pink. I love pink!

PERSON 11: I guess my favorite color is gray. Yes, gray.

PERSON 12: White. My favorite color is white.

PERSON 13: My favorite color? I don't know. Maybe it's gold.

PERSON 3: Hmm . . . my favorite color. I've got it. It's brown! My favorite color is brown! Phew!

42:51 I'M LOOKING FOR A JACKET

SALESPERSON 1: May I help you?

CUSTOMER 1: Yes, please. I'm looking for a jacket.

SALESPERSON 1: Here's a nice jacket.

CUSTOMER 1: But this is a purple jacket!

SALESPERSON 1: That's okay. Purple jackets are very popular this year.

43:20 I'M LOOKING FOR A PAIR OF GLOVES

SALESPERSON 2: Can I help you?

CUSTOMER 2: Yes, please. I'm looking for a pair of gloves.

SALESPERSON 2: Here's a nice pair of gloves.

CUSTOMER 2: But these are green gloves!

SALESPERSON 2: That's okay. Green gloves are very popular this year.

43:37 CLYDE'S CLOTHING

CLYDE: Hello. This is Clyde Clifford, owner of Clyde's Clothing. All of us here at Clyde's Clothing are very proud of our store. At Clyde's, there's clothing for everyone. Are you looking for a hat? Or maybe a belt? How about a sweater? What about an umbrella for those rainy days? Are you looking for a necklace for a special person? What about a beautiful vinyl briefcase? Well, come on over to Clyde's, where there are lots of hats, belts, sweaters, umbrellas, necklaces, and briefcases. So take it from me, Clyde Clifford. At Clyde's, there's clothing for everyone!

JINGLE: If your clothes are old, And you look a mess, Come to Clyde's And dress for less. Clyde's Clothing.

GRAMMAR

Singular/Plural

[s]
> I'm looking for **a** jacket.
> Purple jacket**s** are very popular this year.

[z]
> I'm looking for **an** umbrella.
> Purple umbrella**s** are very popular this year.

[ɪz]
> I'm looking for **a** dress.
> Purple dress**es** are very popular this year.

I'm looking for	a	jacket.
	a pair of	gloves.

Adjectives

> This is a purple jacket.
> These are green gloves.

FUNCTIONS

Offering to Help

May I help you?
Can I help you?

Expressing Want-Desire

I'm looking for *a jacket.*
I need *an umbrella.*

Describing

Here's a nice *jacket.*

But this is a *purple* jacket!
But these are *green* gloves!

Attracting Attention

Hello.

Excuse me.
Pardon me.

Expressing Surprise-Disbelief

But this is a PURPLE jacket!

Giving Location

Shirts are over there/on that
    counter/right over
    there/around the corner.

Greeting People

Hello. How are you today?
    Fine, thank you.
Hi!

Expressing Gratitude

Thank you.
Thank you very much.
Thanks very much.
Thanks a lot.
    You're welcome.
    You're very welcome.

Checking Understanding

A dress?

Expressing Certainty

It's definitely *blue.*

I suppose it's *silver.*

I've got it. It's *brown.*

Expressing Uncertainty

I'm sorry. I don't know.
I just don't know.

I guess *my favorite color is
    gray.*

Maybe *it's gold.*

Asking for Repetition

What's the question again?

- **Clothing**
- **This/That/These/Those**

♪ *"This and that and these and those are my books, supplies, and clothes . . . They're all Side by Side."*

PROGRAM LISTINGS

SBS-TV Backstage Bulletin Board

SBS-TV

TO: Production Crew
Sets and props for this segment:

Office
jackets

Store
counter
gloves

Lost and Found Counter
boots
umbrella

Street Corner
briefcase
calculator
gloves
pen
pencils
photograph
running shorts
umbrella

Laundromat
hat
jacket
shirt
skirt
sweaters

TO: Cast Members
Key words in this segment:

this
that

these
those

45:05 EXCUSE ME. I THINK THAT'S MY JACKET

SOUND CHECK

A. Excuse me. I think this is / that's ¹ my jacket.

B. I don't think so. I think this is / that's ² MY jacket.

A. Oh. You're right. I guess I made a mistake.

A. Excuse me. I think these / those ³ are my gloves.

B. I don't think so. I think these / those ⁴ are MY gloves.

A. Oh. You're right. I guess I made a mistake.

45:47 LOST AND FOUND

SOUND CHECK

A. Is this / that ¹ your umbrella?

B. No, it isn't.

A. Are you sure?

B. Yes. This / That ² umbrella is brown, and my umbrella is black.

A. Hmm. Let me see.

A. Are this / these ³ your boots?

B. No, they aren't.

A. Are you sure?

B. Yes. These / Those ⁴ boots are dirty, and my boots are clean.

I'M TERRIBLY SORRY!

PROP DEPARTMENT

Help the production crew put together the props for this segment.

a. briefcase	c. calculator	e. umbrella	g. running shorts
b. pen	d. gloves	f. pencils	

1 _e_

2 ____

3 ____

4 ____

5 ____

6 ____

7 ____

EDITING MIX-UP

The video editor made a mistake! Put the following lines in the correct order.

____ I'm fine. Oh, my goodness! Look at this mess!

_1_ Oh, I'm terribly sorry!

____ Yes. I'm okay. How about you?

____ Are you okay?

____ Here. Let me help you.

SOUND CHECK

A. I think $\left\{\begin{array}{c}\text{this}\\\text{that}\end{array}\right\}^1$ is your briefcase.

B. No, $\left\{\begin{array}{c}\text{this}\\\text{that}\end{array}\right\}^2$ isn't MY briefcase. $\left\{\begin{array}{c}\text{This is}\\\text{That's}\end{array}\right\}^3$ YOUR briefcase.

$\left\{\begin{array}{c}\text{This}\\\text{That}\end{array}\right\}^4$ is MY briefcase.

A. You're right.

(continued)

B. Is this that [5] your pen?

A. Yes, it is. Thanks.

Is this that [6] your calculator?

B. Yes. Thank you.

A. I think these those [7] are my gloves.

B. Hmm. These Those [8] are nice gloves.

A. Thanks.

B. And this is that's [9] my umbrella.

A. Hmm. Are you sure? I think this is that's [10] MY umbrella.

How about this that [11] umbrella?

B. This is That's [12] definitely not my umbrella. This is That's [13] my umbrella.

A. You're right. And this is that's [14] MY umbrella.

B. Thanks.

A. Thanks.

B. Are these those [15] your pencils?

A. Yes. These Those [16] are my pencils.

Thanks.

A. Is this that [17] your photograph?

B. Yes. It is. These Those [18] are my children.

A. Cute kids!

B. Thank you.

B. And I guess these those [19] are your running shorts.

A. Thanks.

WHOSE THINGS?

Which items in this scene belong to each of these people?

BUSINESSMAN: _briefcase, pen_

BUSINESSWOMAN: _briefcase_

Circle the correct response.

1 Are you okay?

   (a.) Yes. I'm okay.

   b. Thanks.

2 THIS is my briefcase.

   a. Here! Let me help you.

   b. You're right.

3 Is this your pen?

   a. Yes, it is.

   b. This is MY pen.

4 And THAT's my umbrella.

   a. It's okay.

   b. Hmm. Are you sure?

5 Are these your pencils?

   a. Yes. These are my pencils.

   b. Yes. Those are my pencils.

6 Well, I'm really sorry about this.

   a. I'm fine.

   b. No problem.

WRITE THE SCRIPT!

Two other people just bumped into each other! Complete the script any way you wish and then practice it with a friend.

A. Oh, I'm terribly sorry!

B. Are you okay?

A. Yes. I'm okay. How about you?

B. I'm fine. Oh, my goodness! Look at this mess!

A. Here. Let me help you. I think this is your ....

B. No. That isn't MY .... That's YOUR ....

A. You're right.

B. Are these your .... ?

A. Yes, they are. Thanks.

B. And I think this/that/these/those ....

A. ....

B. ....

A. ....

B. ....

Are they saying **this, that, these,** or **those?** It depends on how close the objects are to the person. See if you can tell based on the pictures.

a. I think this is your briefcase.
b. I think that's your briefcase.

a. THAT's my briefcase.
b. THIS is my briefcase.

1

2

a. Is this your pen?
b. Is that your pen?

a. Is that your calculator?
b. Is this your calculator?

3

4

a. I think those are my gloves.
b. I think these are my gloves.

5

a. And THAT's my umbrella.
b. And THIS is my umbrella.

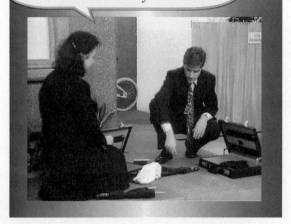

6

a. I think this is MY umbrella.
b. I think that's MY umbrella.

7

a. That is definitely not my umbrella.
b. This is definitely not my umbrella.

8

a. Are these your pencils?
b. Are those your pencils?

9

a. And I guess these are your running shorts.
b. And I guess those are your running shorts.

10

FINISH THE SONG!

hat	those	shirt	suits	that	are
skirt	That's	boots	these	this	

Is ___this___ [1] your sweater?

Is _____ [2] your _____ [3]?

_____ [4] my blue jacket.

That's my pink _____ [5].

_____ [10] and that.

At the laundromat.

This and _____ [11].

At the laundromat.

_____ [12] and _____ [13].

At the laundromat.

Are _____ [14] your mittens?

_____ [15] these your _____ [16]?

_____ [17] are my socks.

Those are my bathing _____ [18].

_____ [22] and _____ [23].

Washing all our clothes.

_____ [24] and _____ [25].

I think _____ [6] is my new _____ [7].

We're looking for _____ [8] and _____ [9].

We're washing all our clothes at the laundromat.

Where _____ [19] my pantyhose?

We're looking for _____ [20] and _____ [21].

We're washing all our clothes at the laundromat.

At the laundromat.

At the laundromat.

(Hey! Give me _____ [26]!)

At the laundromat!

RHYME TIME

Find the rhyming words in the song.

1 Miss _____this_____

2 fat _____

3 nose _____

4 true _____

5 shirt _____

6 cooking _____

THE NEXT LINE

Circle the best response.

1 Excuse me. Is this your jacket?

  (a.) No, that's my jacket over there.

  b. No, thank you.

2 Tell me, are those your children?

  a. Yes, it is.

  b. Yes, they are.

3 Is that your car?

  a. No, it isn't.

  b. No, there isn't.

4 I think this is my coat.

  a. Oh. You're right.

  b. Is this my coat?

5 Are these your boots?

  a. Yes, they are.

  b. Yes, there are.

6 This is my calculator.

  a. Are you sure?

  b. Are you okay?

7 I think those are my glasses.

  a. Oh. You're right.

  b. Are these my glasses?

8 Excuse me. Is this your camera?

  a. Oh, my goodness. Thank you.

  b. No, thank you.

SCRAMBLED SOUND TRACK

The sound track is all mixed up. Put the words in the correct order and then practice the conversation with a friend.

A. `Good` `?` `help` `morning` `I` `Can` `you` `.`   <u>Good morning. Can I help you?</u>

B. `Yes` `for` `.` `looking` `.` `my` `jacket` `I'm`

A. `color` `is` `?` `What` `it`

B. `.` `brown` `It's`

A. `this` `?` `Is` `jacket` `your`

B. `No` `jacket` `That` `.` `.` `my` `isn't`

A. `?` `about` `this` `How` `jacket`

B. `That's` `jacket` `!` `my` `Yes` `very` `.` `.`

  `Thank` `much` `you`

45:05 EXCUSE ME. I THINK THAT'S MY JACKET

PERSON 1: Excuse me. I think that's my jacket.

PERSON 2: Hmm. I don't think so. I think this is MY jacket.

PERSON 1: Oh. You're right. I guess I made a mistake.

PERSON 3: Excuse me. I think those are my gloves.

PERSON 4: Hmm. I don't think so. I think these are MY gloves.

PERSON 3: Oh. You're right. I guess I made a mistake.

45:47 LOST AND FOUND

ATTENDANT: Is this your umbrella?

MAN: No, it isn't.

ATTENDANT: Are you sure?

MAN: Yes. That umbrella is brown, and my umbrella is black.

ATTENDANT: Hmm. Let me see.

ATTENDANT: Are these your boots?

WOMAN: No, they aren't.

ATTENDANT: Are you sure?

WOMAN: Yes. Those boots are dirty, and my boots are clean.

46:21 I'M TERRIBLY SORRY!

BUSINESSMAN: Oh, I'm terribly sorry!

BUSINESSWOMAN: Are you okay?

BUSINESSMAN: Yes. I'm okay. How about you?

BUSINESSWOMAN: I'm fine. Oh, my goodness! Look at this mess!

BUSINESSMAN: Here. Let me help you. I think this is your briefcase.

BUSINESSWOMAN: No, that isn't MY briefcase. That's YOUR briefcase. THIS is MY briefcase.

BUSINESSMAN: You're right.

BUSINESSWOMAN: Is this your pen?

BUSINESSMAN: Yes, it is. Thanks. Is this your calculator?

BUSINESSWOMAN: Yes. Thank you.

BUSINESSMAN: I think those are my gloves.

BUSINESSWOMAN: Hmm. These are nice gloves.

BUSINESSMAN: Thanks.

BUSINESSWOMAN: And THAT's my umbrella.

BUSINESSMAN: Hmm. Are you sure? I think this is MY umbrella. How about THAT umbrella?

BUSINESSWOMAN: This is definitely not my umbrella. THAT's my umbrella.

BUSINESSMAN: You're right. And that's MY umbrella.

BUSINESSWOMAN: Thanks.

BUSINESSMAN: Thanks.

BUSINESSWOMAN:	Are these your pencils?
BUSINESSMAN:	Yes. Those are my pencils. Thanks.
	Is this your photograph?
BUSINESSWOMAN:	Yes. It is. These are my children.
BUSINESSMAN:	Cute kids!
BUSINESSWOMAN:	Thank you.
	And I guess these are your running shorts.
BUSINESSMAN:	Thanks.
	Well, I'm really sorry about this.
BUSINESSWOMAN:	It's okay. No problem.
	Nice bumping into you!

48:18 AT THE LAUNDROMAT—
Music Video

Is this your sweater?
Is this your shirt?
That's my blue jacket.
That's my pink skirt.

I think this is my new hat.
We're looking for this and that.
We're washing all our clothes at the
laundromat.

This and that.
At the laundromat.
This and that.
At the laundromat.
This and that.
At the laundromat.

Are these your mittens?
Are these your boots?
Those are my socks.
Those are my bathing suits.

Where are my pantyhose?
We're looking for these and those.
We're washing all our clothes at the
laundromat.

This and that.
Washing all our clothes.
These and those.
At the laundromat.
At the laundromat.
(Hey! Give me that!)
At the laundromat!

GRAMMAR

This/That

Is **this** your umbrella? Is **that** your car?
This umbrella is brown. **That** car is old.

These/Those

Are **these** your boots? Are **those** your pens?
These boots are old. **Those** pens are new.

FUNCTIONS

Describing

That *umbrella* is *brown*.
Those *boots* are *dirty*.

Asking for and Reporting Information

Is this your *umbrella*?
  No, it isn't.
Are these your *boots*?
  No, they aren't.

Inquiring about Certainty

Are you sure?

Expressing Certainty

I think *that's my jacket.*

Apologizing

I'm sorry.
I'm terribly sorry.

Admitting an Error

I guess I made a mistake.

Expressing Disagreement

I don't think so.

Offering Help

Let me help you!

Expressing Agreement

You're right.

Attracting Attention

Excuse me.

SEGMENT 13

- **Daily Activities**
- **Languages and Nationalities**
- **Simple Present Tense**

"Different cultures, different faces, different foods, and different places . . . Side by Side."

PROGRAM LISTINGS

50:05 PEOPLE AROUND THE WORLD
Host Robert Lynch reports about people and their lives in different parts of the world. In this segment: Antonio from Rome, and Boris and Natasha from Moscow.

52:30 COME TO MEXICO!
Miguel invites viewers to visit Mexico.

53:15 PEOPLE AROUND THE WORLD
Robert Lynch interviews Anna from Athens.

SBS-TV Backstage Bulletin Board

TO: Production Crew
Sets and props for this segment:

TV Studio	Restaurant
globe	table
armchair	flowers
	newspaper
	Mexican food

TO: Cast Members
Key words in this segment:

do	beautiful
drink	country
eat	food
listen	language
live	music
read	newspaper
sing	people
speak	song
	tea
Greek	vodka
Italian	wine
Mexican	wonderful
Russian	around the world
	See you soon.

SOUND CHECK

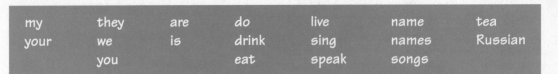

my	they	are	do	live	name	tea
your	we	is	drink	sing	names	Russian
	you		eat	speak	songs	

A. What's _____your_____ ¹ name?

B. _____ ² name _____ ³ Antonio.

A. Where do you _____ ⁴?

B. I _____ ⁵ in Rome.

A. What language do you _____ ⁶?

B. I _____ ⁷ Italian.

A. Tell me, what do you _____ ⁸ every day?

B. I _____ ⁹ Italian food, I _____ ¹⁰ Italian
    wine, and I _____ ¹¹ Italian songs.

A. What _____ ¹² your names?

B. My name is Boris.

C. And _____ _____ _____ ¹³ Natasha.

A. Where _____ _____ ¹⁴ live?

C. We _____ ¹⁵ in Moscow.

A. What language _____ ¹⁶ you _____ ¹⁷?

B. _____ _____ ¹⁸ Russian.

A. Boris and Natasha, tell our audience:
    What _____ ¹⁹ you _____ ²⁰ every day?

B. We _____ ²¹ Russian food. We _____ ²² Russian vodka.

C. YOU drink Russian vodka. I drink Russian _____ ²³.

B. And we _____ ²⁴ Russian songs.

A. Yes, ladies and gentlemen, their

_____ _____ 25 Boris

and Natasha. _____

_____ 26 here in Moscow!

_____ 27 speak _____ 28!

They _____ 29 Russian food!

They _____ 30 Russian vodka!

Don't forget the Russian _____ 31!

They _____ 32 Russian tea!

And they sing Russian _____ 33!

CLOSE-UP

Robert Lynch is interviewing YOU on "People Around the World!"

What's your name?

....

Where do you live?

....

What language do you speak?

....

And what do you do every day?

....
....

You're Robert Lynch! Interview the famous singer, Slim Wilkins.

1	_____What's your name_____?	My name is Slim Wilkins.
2	_____?	I speak English.
3	_____?	I live in Nashville.
4	_____ day?	I play the guitar and sing.

`52:30` # COME TO MEXICO!

SOUND CHECK

A. What's $\begin{matrix} her \\ his \end{matrix}$ ¹ name?

B. $\begin{matrix} Her \\ His \end{matrix}$ ² name $\begin{matrix} is \\ his \end{matrix}$ ³ Miguel.

A. Where $\begin{matrix} does \\ do \end{matrix}$ ⁴ he $\begin{matrix} live \\ lives \end{matrix}$ ⁵?

B. He $\begin{matrix} live \\ lives \end{matrix}$ ⁶ in Mexico City.

A. What language $\begin{matrix} does \\ do \end{matrix}$ ⁷ he $\begin{matrix} speak \\ speaks \end{matrix}$ ⁸?

B. He $\begin{matrix} speak \\ speaks \end{matrix}$ ⁹ Spanish.

A. What $\begin{matrix} does \\ do \end{matrix}$ ¹⁰ he $\begin{matrix} does \\ do \end{matrix}$ ¹¹ every day?

B. He $\begin{matrix} eat \\ eats \end{matrix}$ ¹² Mexican food, he $\begin{matrix} read \\ reads \end{matrix}$ ¹³ Mexican newspapers,

and he $\begin{matrix} listen \\ listens \end{matrix}$ ¹⁴ to Mexican music.

A COMMERCIAL FOR YOUR COUNTRY

Using the script below, write a commercial for YOUR country and present it on Side by Side TV.

Hello. My name is .... I live here in
                              (your name)

.... You know, ... is a wonderful
  (your city)   (your country)

country. Come to .... and meet the people, eat the food, and
                 (your country)

listen to our beautiful music. Come see the .... !

Don't forget the .... ! In ....
                                   (your country)

there are wonderful .... s, ... s,

and .... s! So visit .... ! See you soon!
                                   (your country)

PEOPLE AROUND THE WORLD

WHO IS SHE?

ATHENS

My name is Anna.
I **live** here in Athens.
I **speak** Greek.
I **eat** Greek food,
I **read** Greek newspapers, and
I **listen** to Greek music.

1 What's her name?

2 Where does she live?

3 What language does she speak?

4 What does she do every day?

Her name ___is___ Anna.

_____ in Athens.

_____ Greek.

_____ Greek food,

_____ Greek

newspapers, and _____
to Greek music.

AN INTERVIEW

Interview a friend.

1 _____ you live?   I _____

2 _____ speak?   I _____

3 _____ do every day? I _____

Now tell about your friend.

1 He/She _____.

2 He/She _____.

3 He/She _____.

WRONG LINE

Cross out the mistake.

1. ~~Jack~~
    ~~Marie~~   lives in Montreal.
    ~~Jack and Marie~~

4. ~~The children~~
    ~~The teacher~~   speak English at school.
    ~~I~~

2. ~~We~~
    ~~Peter~~   play baseball every day.
    ~~They~~

5. ~~My Husband~~
    ~~Sam and I~~   listens to the radio every day.
    ~~She~~

3. What does   ~~Jane~~
                ~~you~~   do every day?
                ~~Mark~~

6. What language do   ~~you~~
                      ~~they~~   speak?
                      ~~Bob~~

SCRAMBLED SOUND TRACK

The sound track is all mixed up. Put the words in the correct order.

1. | Moscow | here | . | live | in | I |

    I live here in Moscow.

2. | language | speak | they | do | What | ? |

    _____

3. | Where | ? | live | Mrs. | Johnson | and | Mr. | do |

    _____

4. | He | wine | eats | Greek | and | . | food | drinks | Greek | he |

    _____

5. | do | your | every | does | teacher | ? | What | day |

    _____

AROUND THE WORLD

1. What's her name?   Her name is Fifi.

2. _____ she live?   _____

3. _____ do every day?   She _____

`50:05` PEOPLE AROUND THE WORLD

ROBERT LYNCH: Antonio is from Rome! Boris and Natasha are from Moscow! And Anna is from Athens!

Who are these people?! Where do they live?! What languages do they speak?! And what do they do every day?! I'm Robert Lynch! Join me now as we meet *People Around the World!*

ROBERT LYNCH: What's your name?
ANTONIO: My name is Antonio.
ROBERT LYNCH: Where do you live?
ANTONIO: I live in Rome.
ROBERT LYNCH: What language do you speak?
ANTONIO: I speak Italian.
ROBERT LYNCH: Tell me, what do you do every day?
ANTONIO: I eat Italian food, I drink Italian wine, and I sing Italian songs.

ROBERT LYNCH: What are your names?
BORIS: My name is Boris.

NATASHA: And my name is Natasha.
ROBERT LYNCH: Where do you live?
NATASHA: We live in Moscow.
ROBERT LYNCH: What language do you speak?
BORIS: We speak Russian.
ROBERT LYNCH: Boris and Natasha, tell our audience: What do you do every day?
NATASHA: We eat Russian food.
BORIS: We drink Russian vodka.
NATASHA: YOU drink Russian vodka. I drink Russian tea.
BORIS: And we sing Russian songs.
ROBERT LYNCH: Yes, ladies and gentlemen, their names are Boris and Natasha! They live here in Moscow! They speak Russian! They eat Russian food! They drink Russian vodka!
NATASHA: Don't forget the Russian tea!
ROBERT LYNCH: They drink Russian tea! And they sing Russian songs!

And what about you? What's YOUR name? Where do you live? What language do you speak? And what do you do every day? We'll be right back with more *People Around the World!*

`52:30` COME TO MEXICO!

`53:15` PEOPLE AROUND THE
WORLD—Continued

PERSON 1: What's his name?
PERSON 2: His name is Miguel.
PERSON 1: Where does he live?
PERSON 2: He lives in Mexico City.
PERSON 1: What language does he speak?
PERSON 2: He speaks Spanish.
PERSON 1: What does he do every day?
PERSON 2: He eats Mexican food, he reads Mexican newspapers, and he listens to Mexican music.
MIGUEL: ¡Buenos días! My name is Miguel. I live here in Mexico City. You know, Mexico is a wonderful country. Come to Mexico and meet the people, eat the food, and listen to our beautiful music. See you soon?

ROBERT LYNCH: And now, *People Around the World* takes you to Greece! What's your name?
ANNA: My name is Anna.
ROBERT LYNCH: Where do you live, Anna?
ANNA: I live here in Athens.
ROBERT LYNCH: What language do you speak?
ANNA: I speak Greek.
ROBERT LYNCH: Tell our viewers, Anna: What do you do every day?
ANNA: I eat Greek food, I read Greek newspapers, and I listen to Greek music.
ROBERT LYNCH: Yes, ladies and gentlemen, her name is Anna! She lives in Athens! She speaks Greek! She eats Greek food! She reads Greek newspapers! And she listens to Greek music! Thank you, Anna.

And that's our program for today! This is Robert Lynch, inviting you to join us next time for another edition of *People Around the World!*

GRAMMAR

Simple Present Tense

Where	do	I we you they	live?
	does	he she it	

I We You They	live	in Rome.
He She It	lives	

FUNCTIONS

Asking for and Reporting Information

What's your name?
   My name is *Antonio.*
Where do you live?
   I live in *Rome.*
What language do you speak?
   I speak *Italian.*
What do you do every day?
   I eat *Italian food.*

Tell *me/our audience/our viewers,* _____?

You know, *Mexico is a wonderful country.*

Leave Taking

See you soon.

ANSWER KEY ●●

SEGMENT 1

Page 2

SOUND CHECK
1. name   name
2. address   address
3. phone   phone
   number   number
4. from   from

WHOSE LINE?
1. Maria
2. Teacher
3. Maria
4. Maria
5. Teacher

Page 3

SOUND CHECK
1. a   6. b
2. b   7. b
3. b   8. a
4. a   9. a
5. b

Page 4

ON CAMERA
Name: William Chen
Address: 694 River Street
           Brooklyn, New York
Telephone: 469-7750
Social Security
   Number: 044-35-9862

Page 5

EDITING MIX-UP
   3
   6
   5
   1
   4
   2

1. William Chen.
2. 694 River Street.
3. Yes.
4. 469-7750.
5. 044-35-9862.
6. ....

SOUND CHECK
1. name
2. address
3. Yes
4. telephone, number
5. Social Security
6. zip code

Page 6

SCRAMBLED SOUND TRACK
1. What's your name?
2. What's your telephone number?
3. Where are you from?
4. My address is 842 Main Street.
5. My Social Security number is 105-36-8954.

WHAT'S MY LINE?
1. your   4. are
2. What's   5. your
3. I'm   6. What's

ON CAMERA
1. What's your name?
2. What's your address?
3. What's your telephone number?
4. Where are you from?

SEGMENT 2

Page 10

SETS AND SCENERY
1. kitchen   4. bedroom
2. dining room   5. yard
3. basement   6. living room

SOUND CHECK
1. a
2. b
3. a

SOUND CHECK
1. b
2. a
3. b

Page 11

SOUND CHECK
1. b
2. b
3. a
4. a
5. b

Page 12

WHOSE LINE?
1. Mother   6. Mother
2. Mother   7. Billy
3. Billy   8. Mother
4. Mother   9. Mother
5. Billy   10. Billy

WHERE IS EVERYBODY?
1. He's
2. She's
3. They're
4. I'm

Page 13

EDITING MIX-UP
   3
   10
   12
   6
   2
   7
   1
   11
   8
   4
   5
   13
   9

Page 14

SCRAMBLED SOUND TRACK
1. She's in the living room.
2. Where are you?
3. The car is in the garage.
4. Mrs. Wong is in the kitchen.
5. Where are Paul and Tom?
6. What's your favorite room?
7. Where is Mrs. Ramirez?
8. Fred is in the basement.
9. Is everything okay at home?

WHAT'S MY LINE?
1. is   4. is   7. and
2. am   5. is   8. We're
3. are   6. are

SCRAMBLED WORDS
1. dining room   4. bedroom
2. garage   5. living room
3. yard   6. basement

SEGMENT 3

Page 18

SOUND CHECK
1. He's, restaurant   5. I'm, library
2. She's, bank   6. We're, park
3. They're, supermarket   7. You're, hospital
4. It's, zoo

Page 19

FINISH THE SET!
1. g   5. b
2. e   6. c
3. a   7. d
4. f

Page 20

YES OR NO?
1. Yes   5. Yes   8. No
2. No   6. No   9. No
3. Yes   7. No   10. No
4. Yes

Page 21

WHAT'S MY LINE?
1. Tom   4. I
2. we   5. the car
3. Sue and Jane   6. Rita

WHAT'S THE QUESTION?
1. Where's   5. Where's
2. Where are   6. Where are
3. Where's   7. Where am
4. Where are   8. Where are

SCRAMBLED WORDS
1. garage   4. supermarket
2. kitchen   5. post office
3. library

Page 22

TV CROSSWORD
See page 139.

SEGMENT 4
Page 26

SCRIPT CHECK
1. cooking   4. sleeping
2. watching TV   5. reading
3. eating   6. studying

WHAT'S HAPPENING?
1. a   4. b
2. b   5. a
3. a   6. b

Page 27

FINISH THE RAP!
1. Where's   6. kitchen
2. He's   7. Who's
3. What's   8. in
4. Eating   9. doing
5. lunch

Page 28

WHAT'S THE LINE?
1. b, d, e, h
2. b, d, f, g
3. b, c, f, h

HELLO, EVERYBODY!
1. I'm, kitchen, eating breakfast
2. I'm, park, eating lunch
3. We're, dining room, eating dinner

Page 29

FINISH THE RAP!
1. Where's   5. reading
2. bedroom   6. in
3. she   7. Who's
4. book   8. What's

Page 30

YES OR NO?
1. Yes   5. No
2. No   6. Yes
3. Yes   7. No
4. Yes   8. No

Page 31

FINISH THE RAP!
1. are   7. kitchen
2. They're   8. Where's
3. are   9. He's
4. Watching   10. What's
5. bedroom   11. lunch
6. in

YES OR NO?
1. No   5. Yes
2. No   6. No
3. Yes   7. No
4. Yes

Page 32

FINISH THE SCRIPT!
1. shining   7. studying
2. yard   8. kitchen
3. reading   9. cooking
4. living room   10. playing
5. listening   11. bedroom
6. dining room

SCRAMBLED WORDS
1. eating   5. dancing
2. listening   6. playing
3. watching   7. doing
4. reading   8. cooking

Page 33

SCRAMBLED SOUND TRACK
1. What are they doing?
2. What's Sally doing?
3. Luis is eating lunch in the kitchen.
4. They're watching TV in the living room.
5. I'm reading a book in the bedroom.
6. Where are Peter and Tommy?
7. Antonio is cooking dinner in the kitchen.
8. Mariko is studying in her bedroom.

WHAT'S MY LINE?
1. is   5. are
2. are   6. is
3. is   7. are
4. are   8. are

SEGMENT 5
Page 38

SOUND CHECK 1
1. a, c
2. a, b
3. b, c
4. b, c
5. a, b

SOUND CHECK 2
1. I'm, my
2. He's, his
3. She's, her
4. We're, our
5. They're, their

Page 39

THE WRONG CAPTIONS
1. a, c
2. a, b
3. b, c
4. a, c
5. a, b

SOUND CHECK
1. washing   8. her
2. hair   9. Are
3. is   10. are
4. He's   11. our
5. his   12. They're
6. she   13. their
7. She's

Page 40

WHAT'S EVERYBODY DOING?
1. a   5. a
2. b   6. a
3. b   7. b
4. b   8. b

EDITING MIX-UP 1
4
3
1
8
2
5
7
6

Page 41

EDITING MIX-UP 2

8
2
4
6
1
7
5
3

MATCH THE LINES!
1. e   4. b
2. d   5. c
3. a

Page 42

WRONG LINES
1. sink   5. hair
2. homework   6. playing
3. TV   7. kitchen
4. reading   8. drinking

WHAT ARE THEY SAYING?
1. Yes, they are.   4. Yes, she is.
2. Yes, I am./   5. Yes, I am.
   Yes, we are.   6. Yes, it is.
3. Yes, he is.

Page 43

TV CROSSWORD
See page 139.

SEGMENT 6

Page 48

SOUND CHECK
1. tall   3. Is
2. He's   4. short

WHICH CAPTION?
1. tall   8. ugly
2. short   9. rich
3. young   10. poor
4. old   11. large
5. heavy   12. small
6. thin   13. noisy
7. handsome   14. quiet

Page 49

YES, NO, OR MAYBE?
1. No   6. Maybe
2. Yes   7. Maybe
3. Maybe   8. No
4. No   9. Yes
5. No

Page 50

SCENE CHECK
1. Neil   4. old
2. mother   5. noisy
3. large   6. difficult

WHOSE LINE?
1. Mother   7. Mother
2. Mother   8. Neil
3. Neil   9. Neil
4. Mother   10. Mother
5. Neil   11. Mother
6. Neil   12. Neil

Page 51

THE NEXT LINE
1. a   3. a   5. a
2. b   4. a   6. b

A LETTER FROM NEIL
1. large   5. quiet
2. small   6. difficult
3. cheap   7. interesting
4. noisy   8. happy

Page 52

GETTING TO KNOW THE PLAYERS
1. Rich   4. Long
2. Young   5. Little
3. Long   6. Little

WHOSE LINE?
1. Announcer   6. Rich Young
2. Rich Young   7. Rich Young
3. Rich Young   8. Larry Little
4. Lisa Long   9. Rich Young
5. Rich Young 10. Larry Little

Page 53

OPPOSITE ADJECTIVES
1. old (Larry)   5. little (Lisa)
2. ugly (Lisa)   6. heavy (Larry)
3. quiet (Lisa)   7. expensive (Lisa)
4. single (Larry)

THE BONUS ROUND!
1. tall   Yes
2. rich   Yes
3. large   Yes
4. beautiful   Yes
5. thin   No
6. easy   Yes
7. married   Yes

Page 54

WHAT'S THAT ADJECTIVE?
1. old   5. difficult
2. large   6. short
3. expensive   7. young
4. single   8. quiet

TRUE OR FALSE?
1. No, it isn't.   5. Yes, it is.
2. No, it isn't.   6. No, it isn't.
3. Yes, they are.   7. Yes, it is.
4. No, it isn't.   8. ....

SEGMENT 7

Page 60

WEATHER CHECK
1. It's sunny.   5. It's hot.
2. It's cloudy.   6. It's warm.
3. It's raining.   7. It's cool.
4. It's snowing.   8. It's cold.

YES OR NO?
1. No, it isn't.   4. No, it isn't.
2. Yes, it is.   5. Yes, it is.
3. No, it isn't.   6. Yes, it is.

Page 62

YES OR NO?
1. Yes   5. Yes
2. Yes   6. No
3. No   7. Yes
4. No

THE NEXT LINE
1. a   4. a
2. b   5. a
3. b   6. b

Page 64

WEATHER CHECK
1. sunny
2. cloudy
3. raining
4. snowing
5. hot
6. warm
7. cool
8. cold

WEATHER RECAP
1. It's sunny   5. It's hot
2. It's cloudy   6. It's warm
3. It's raining   7. It's cool
4. It's snowing   8. It's cold

WEATHER CHALLENGE!
1. hot   9. hot
2. raining   10. raining
3. warm   11. sunny
4. snowing   12. cool
5. cool   13. snowing
6. cold   14. cold
7. cloudy   15. warm
8. snowing

SEGMENT 8

Page 70

PREVIEW

1. wife
2. husband
3. son
4. daughter
5. brother
6. sister
7. parents
8. mother
9. father
10. aunt
11. uncle
12. cousin
13. grandparents
14. grandmother
15. grandfather

Page 71

EDITING MIX-UP

6
4
1
7
3
9
5
2
8
10

WHAT'S THE RESPONSE?

1. d
2. c
3. a
4. e
5. b

ALL IN THE FAMILY

1. aunt
2. uncle
3. grandmother
4. grandfather
5. brother
6. sister
7. wife, husband
8. son, daughter

Pages 72–73

SOUND CHECK

1. mother
2. playing
3. playing
4. living room
5. father
6. eating
7. dinner
8. restaurant
9. daugher
10. son
11. playing
12. yard
13. sister
14. brother
15. singing
16. They're
17. singing
18. husband
19. husband
20. watching
21. sleeping
22. brother
23. wife
24. playing
25. playing
26. grandmother
27. grandfather
28. aunt
29. uncle
30. cousin
31. grandmother

Page 74

INSTANT REPLAY

1. c
2. c
3. c
4. a
5. b
6. b
7. a

EDITING MIX-UP

1. 2
   1
2. 2
   1
3. 1
   2
4. 1
   2
5. 2
   1
6. 2
   1

Page 75

FINISH THE SONG!

1. smiling
2. living
3. working
4. looking
5. hanging
6. dancing
7. having
8. crying
9. looking
10. hanging
11. smiling
12. Looking

Page 76

IS IT TRUE?

1. False
2. False
3. True
4. False
5. False
6. True
7. False
8. True
9. False

RHYME TIME

1. Dad
2. hall
3. far
4. son
5. at
6. looking
7. away
8. sister
9. Hi
10. brother

I'M ALL MIXED UP!

1. We're having a good time.
2. I'm looking at the photographs.
3. My daughter's working in Detroit.
4. My little sister's crying.
5. It's a very special day.
6. I'm very far away.
7. It's my brother's wedding day.
8. I love you, Dad.
9. My son Robert's married now.
10. I'm so happy!

Page 77

SCRAMBLED WORDS

1. sister
2. grandmother
3. daughter
4. brother
5. son
6. aunt
7. uncle
8. cousin

SEGMENT 9

Page 82

SOUND CHECK

1. next to
2. across from
3. between
4. around the corner from

Page 83

PICTURE THIS

b.

SOUND CHECK

1. there
2. in
3. neighborhood
4. There's
5. on
6. next to

Page 84

COMPLETE THE NEIGHBORHOOD MAP

See page 140.

DESCRIBE THE NEIGHBORHOOD

1. around the corner from the movie theater
2. across from the bakery
3. next to the police station
4. between the school and the movie theater
5. across from the police station

Page 85

SOUND CHECK

1. across from
2. school
3. there
4. There's
5. is
6. next to
7. post office
8. on
9. am
10. around
11. a
12. neighborhood
13. There's
14. Where
15. across from
16. Is

Page 86

WHAT'S MY LINE?

1. there
2. there's
3. to
4. a
5. nearby
6. on
7. from

WHAT ARE THEY SAYING?

1. next to
2. between
3. next to, school
4. around the corner from
5. Where's the movie theater?
6. It's on State Street, next to the park./It's around the corner from the bakery.

SEGMENT 10

Page 90

SCRIPT CHECK

1. table
2. window
3. closet
4. elevator
5. stove
6. jacuzzi
7. refrigerator
8. washing machine

YES OR NO?

1. Yes
2. No
3. No
4. Yes
5. No
6. No
7. No

Page 91

WHAT'S THE LINE?

1. b, d
2. a, d, e

WHAT'S THE LINE?

c, e, g

WHICH ONE?

✓ Apartment B

Page 92

THE NEXT LINE

1. b
2. b
3. a
4. b
5. a
6. b
7. a
8. b

Page 93

FINISH THE RAP!

1. isn't
2. There's
3. There's
4. There
5. are
6. isn't
7. There
8. isn't
9. There
10. aren't
11. There's
12. there
13. are
14. There
15. are

YES OR NO?

1. Yes
2. Yes
3. Yes
4. No
5. Yes
6. Yes
7. No

Page 94

SCRAMBLED WORDS

1. stove
2. closet
3. elevator
4. apartment
5. laundromat
6. table
7. washing machines
8. window
9. jacuzzi
10. refrigerator

WHAT'S THE LINE?

1. j
2. f
3. g
4. l
5. a
6. k
7. d
8. b
9. h
10. c
11. e
12. i

Page 95

SCRAMBLED SOUND TRACK

1. There's a very large living room in the apartment.
2. There isn't an elevator in the building.
3. There are three large windows in the dining room.
4. I'm looking for an apartment to rent.
5. Is there a stove in the apartment?
6. Would you like to see the apartment?

WHAT'S MY LINE?

1. There are
2. Is
3. There's
4. Are
5. There's
6. isn't
7. are
8. Is

SEGMENT 11

Page 100

PROP DEPARTMENT

1. a shirt
2. shirts
3. a coat
4. coats
5. a tie
6. ties
7. an umbrella
8. umbrellas
9. a dress
10. dresses
11. a watch
12. watches

WHOSE LINE?

1. Customer
2. Salesperson
3. Customer
4. Salesperson
5. Salesperson
6. Customer
7. Salesperson
8. Salesperson

Page 101

EDITING MIX-UP

  3
  1
  6
  2
  5
  4

MATCH THE LINES!

1. e
2. a
3. b
4. c
5. d

Page 102

SOUND CHECK

1. a
2. a
3. b
4. b
5. b
6. b
7. b
8. b
9. a
10. b
11. b
12. a
13. a
14. b
15. a

Page 103

SOUND CHECK

1. jacket
2. jacket
3. a
4. jackets

SOUND CHECK

1. gloves
2. pair
3. these
4. gloves

Page 104

PROP DEPARTMENT

1. a pair of earrings
2. a pair of pajamas
3. a pair of shoes
4. a pair of boots
5. a pair of socks
6. a pair of pants

AND NOW A WORD FROM OUR SPONSOR

1. a hat
2. a belt
3. a sweater
4. an umbrella
5. a necklace
6. briefcase
7. hats
8. belts
9. sweaters
10. umbrellas
11. necklaces
12. briefcases

Page 106

WHAT ARE THEY SAYING?

1. a blue jacket
2. a white dress
3. a black umbrella
4. a red tie
5. a, yellow belt
6. a gray coat
7. a pink shirt
8. a gold watch
   a, gold watch

Page 107

WHAT'S MY LINE?

1. an
2. Vinyl briefcases
3. a
4. Watches
5. of
6. This, is
7. pants
8. orange

A CLOTHING STORE FLYER
- HATS
- BELTS
- GLOVES
- DRESSES
- WATCHES
- JACKETS
- TIES

SEGMENT 12

Page 112

SOUND CHECK
1. that's
2. this is
3. those
4. these

SOUND CHECK
1. this
2. That
3. these
4. Those

Page 113

PROP DEPARTMENT
1. e
2. b
3. d
4. c
5. f
6. a
7. g

EDITING MIX-UP

| 4 |
| 1 |
| 3 |
| 2 |
| 5 |

Pages 113–114

SOUND CHECK
1. this
2. that
3. That's
4. This
5. this
6. this
7. those
8. These
9. that's
10. this is
11. that
12. This is
13. That's
14. that's
15. these
16. Those
17. this
18. These
19. these

WHOSE THINGS?

BUSINESSMAN: briefcase, pen, gloves, umbrella, pencils, running shorts

BUSINESSWOMAN: briefcase, calculator, umbrella, photograph

Page 115

THE NEXT LINE
1. a
2. b
3. a
4. b
5. b
6. b

Pages 116–117

WHAT'S THE LINE?
1. a
2. b
3. a
4. b
5. a
6. a
7. a
8. b
9. a
10. a

Page 118

FINISH THE SONG!
1. this
2. this
3. shirt
4. That's
5. skirt
6. this
7. hat
8. this
9. that
10. This
11. that
12. This
13. that
14. these
15. Are
16. boots
17. Those
18. suits
19. are
20. these
21. those
22. This
23. that
24. These
25. those
26. that

RHYME TIME
1. this
2. that, hat, laundromat
3. those, clothes, pantyhose
4. new, blue
5. skirt
6. looking

Page 119

THE NEXT LINE
1. a
2. b
3. a
4. a
5. a
6. a
7. a
8. a

SCRAMBLED SOUND TRACK
A. Good morning. Can I help you?
B. Yes. I'm looking for my jacket.
A. What color is it?
B. It's brown.
A. Is this your jacket?
B. No. That isn't my jacket.
A. How about this jacket?
B. Yes! That's my jacket. Thank you very much.

SEGMENT 13

Pages 124–125

SOUND CHECK
1. your
2. My
3. is
4. live
5. live
6. speak
7. speak
8. do
9. eat
10. drink
11. sing
12. are
13. my name is
14. do you
15. live
16. do
17. speak
18. We speak
19. do
20. do
21. eat
22. drink
23. tea
24. sing
25. names are
26. They live
27. They
28. Russian
29. eat
30. drink
31. tea
32. drink
33. songs

Page 126

ON CAMERA
1. What's your name?
2. What language do you speak?
3. Where do you live?
4. What do you do every day?

SOUND CHECK
1. his
2. His
3. is
4. does
5. live
6. lives
7. does
8. speak
9. speaks
10. does
11. do
12. eats
13. reads
14. listens

Page 128

WHO IS SHE?
1. is
2. She lives
3. She speaks
4. She eats, she reads, she listens

AN INTERVIEW
1. Where do
2. What language do you
3. What do you

Page 129

WRONG LINE
1. Jack and Marie
2. Peter
3. you
4. The teacher
5. Sam and I
6. Bob

SCRAMBLED SOUND TRACK
1. I live here in Moscow.
2. What language do they speak?
3. Where do Mr. and Mrs. Johnson live?
4. He eats Greek food and he drinks Greek wine.
5. What does your teacher do every day?

AROUND THE WORLD
1. Her name is Fifi.
2. Where does She lives in Paris.
3. What does she

Page 22

TV CROSSWORD

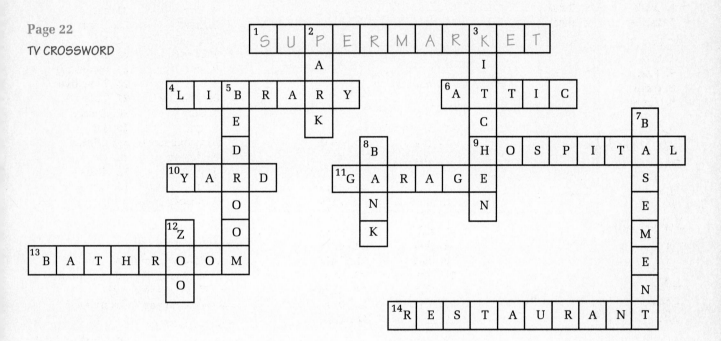

Page 43

TV CROSSWORD

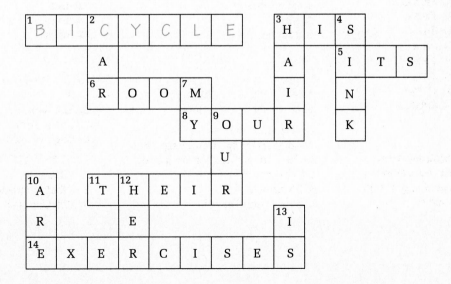

Page 84

COMPLETE THE NEIGHBORHOOD MAP

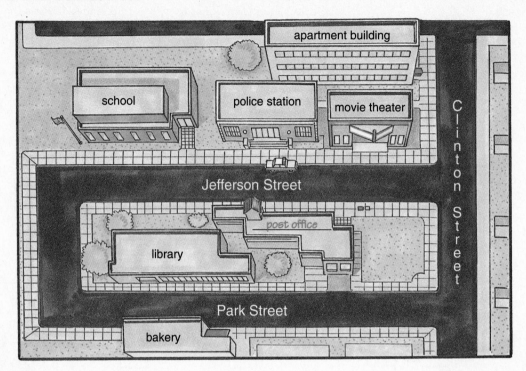